Power Quality Measurement and Analysis Using Higher-Order Statistics

Power Quality Measurement and Analysis Using Higher-Order Statistics

Understanding HOS Contribution on the Smart(er) Grid

Olivia Florencias-Oliveros
Juan-José González-de-la-Rosa
José-María Sierra-Fernández
Manuel-Jesús Espinosa-Gavira
Agustín Agüera-Pérez
José-Carlos Palomares-Salas

University of Cádiz
Department of Automation Engineering, Electronics, Architecture and Computer Networks,
Research Group PAIDI-TIC-168. Computational Instrumentation and Industrial Electronics (ICEI),
Higher Technical School of Engineering of Algeciras (ETSIA), Spain

This edition first published 2023
© 2023 John Wiley & Sons Ltd

The right of Olivia Florencias-Oliveros, Juan-José González-de-la-Rosa, José-María Sierra-Fernández, Manuel-Jesús Espinosa-Gavira, Agustín Agüera-Pérez, and José-Carlos Palomares-Salas to be identified as the author of this work has been asserted in accordance with law.

Registered Offices
John Wiley & Sons Ltd, The Atrium, Southern Gate, Chichester, West Sussex, PO19 8SQ, UK

Editorial Office
The Atrium, Southern Gate, Chichester, West Sussex, PO19 8SQ, UK

For details of our global editorial offices, customer services, and more information about Wiley products visit us at www.wiley.com.

Wiley also publishes its books in a variety of electronic formats and by print-on-demand. Some content that appears in standard print versions of this book may not be available in other formats.

Library of Congress Cataloging-in-Publication Data
Names: Florencias-Oliveros, Olivia, author.
Title: Power quality measurement and analysis using higher-order statistics :
    understanding HOS contribution on the smart(er) grid / Olivia
    Florencias-Oliveros [and five others].
Description: Hoboken, NJ : Wiley, 2023. | Includes bibliographical
    references and index.
Identifiers: LCCN 2022023836 (print) | LCCN 2022023837 (ebook) | ISBN
    9781119747710 (cloth) | ISBN 9781119747765 (adobe pdf) | ISBN
    9781119747741 (epub)
Subjects: LCSH: Electric power systems–Quality control. | Order
    statistics.
Classification: LCC TK1010 .F55 2023 (print) | LCC TK1010 (ebook) | DDC
    621.31–dc23/eng/20220722
LC record available at https://lccn.loc.gov/2022023836
LC ebook record available at https://lccn.loc.gov/2022023837

Cover Design: Wiley
Cover Image: © Pand P Studio/Shutterstock

Set in 9.5/12.5pt STIXTwoText by Straive, Pondicherry, India
Printed and bound by CPI Group (UK) Ltd, Croydon, CR0 4YY

To all the researchers that have inspired this work, those working to bridge the gap between signal analysis and power metering

Contents

Preface

The so-called digital energy networks are gathering numerous elements that have emerged from different branches of Engineering and Science. Concepts such as Internet of Things (IoT), Big Data, Smart Cities, Smart Grid and Industry 4.0 all converge together with the goal of working more efficiently, and this fact inevitably leads to Power Quality (PQ) assurance. Apart from its economic losses, a bad PQ implies serious risks for machines and consequently for people. Many researchers are endeavouring to develop new analysis techniques, instruments, measurement methods and new indices and norms that match and fulfil requirements regarding the current operation of the electrical network. This book offers a compilation of the recent advances in this field. The chapters range from computing issues to technological implementations, going through event detection strategies and new indices and measurement methods that contribute significantly to the advance of PQ analysis. Experiments have been developed within the frames of research units and projects and deal with real data from industry and public buildings. Human beings have an unavoidable commitment to sustainability, which implies adapting PQ monitoring techniques to our dynamic world, defining a digital and smart concept of quality for electricity.

PQ analysis is evolving continuously, mainly due to the incessant growth and development of the smart grid (SG) and the incipient Industry 4.0, which demands quick and accurate tracking of the electrical power dynamics. Much effort has been put on two main issues. First, numerous distributed energy resources and loads provoke highly fluctuating demands that alter the ideal power delivery conditions, introducing at the same time new types of electrical disturbances. For this reason, permanent monitoring is needed in order to track this a priori unpredictable behaviour. Second and consequently, the huge amount of data (Big Data) generated by the measurement equipment during a measurement campaign is usually difficult to manage due to different causes, such as complex structures and communication protocols that hinder accessibility to storage units, and the limited possibilities of monitoring equipment, based on regulations that do not reflect the current network operation.

The introduction of new indicators in PQ is one of the main subjects of discussion in the CIRED/CIGRÉ working group; however, it is necessary to solve future challenges from new perspectives. Indeed, this book proposes to spread the use of PQ indices based on HOS from event detection up to cycle-to-cycle continuous monitoring, taking advantage of their most simple calculations in order to detect the effect of multiple loads acting/working together on a node for a specific length of time.

Chapter 1 introduces the State of the Art in the power quality field and will help researchers to bridge the gap between traditional methods and those applications that use HOS analysis.

Chapters 2–5 propose different and experimental approaches that have been used to validate HOS applications in monitoring the power system.

Table A summarizes the monitoring objectives that would be accomplished using HOS as part of the results of this book and according to the

Table A HOS approach related to different applications.

Monitoring objective	Variables	Sampling rate	Data averaging window	Reference
Compliance verification-connections agreements/premium power contracts	Voltage sags or voltage swells	5 Hz	As specified in the contract	Chapter 3
Performance analysis	Steady-state voltage Voltage sags and swells Highest or lowest RMS voltage per 1 (or 10 min) Fundamental frequency deviations	5 Hz	10 min averaging window 1 min averaging window	Chapter 3
Site characterization		20 kHz		Tables Chapter 4 Chapter 5
Troubleshooting	Disturbance depending on the nature of the problem being investigated			Chapter 3 Chapter 4

topics proposed in the Guideline for Selection of Monitoring Parameters. Compared with other simpler methods, such as RMS measurements, HOS are not sensitive to noise. In Chapter 3, the authors demonstrated that HOS can help to detect fundamental frequency changes in the bi-dimensional plane and Chapter 4 introduces techniques in the frequency domain, such as spectral kurtosis.

Overall, here the authors summarize the last 10 years of power quality research based on HOS techniques that would be incorporated in future PQ measurement campaigns, in order to accomplish the monitoring challenges of the next generation of advanced metering infrastructure in terms of compression, as well as reporting PQ efficiently.

This book gathers new advances in techniques and procedures to describe, measure and visualize the behaviour of the electrical supply, from physical instruments to statistical signal processing (SSP) techniques and new indexes for PQ that try to go beyond traditional norms and standards. The authors are recognized experts in the field, committed to a main goal: to provide new instrumental and analytical tools to help mitigate the serious consequences of a bad PQ in our digitized society, and thus enhancing energy efficiency for a more sustainable development.

Olivia Florencias-Oliveros
Juan-José González-de-la-Rosa

About the Authors

Dr Olivia Florencias-Oliveros received a PhD degree in Energy and Sustainable Engineering in 2020 (*summa cum laude*) from the University of Cádiz, Spain. She is a lecturer at the University of Cádiz, in the Research Group in Computational Instrumentation and Industrial Electronics (PAIDI-TIC168), and an IEEE Member in the Power and Energy Society and IEEE Measurement and Instrumentation Society. Her research interests include Energy Technologies to manage Energy Efficiency and Renewable Energies: smart grids, energy monitoring techniques in power systems, power quality, smart metering, computational instrumentation technologies, sensor networks, IoT in smart buildings, big data and HOS statistics.

Dr Juan-José González-de-la-Rosa received an MSc degree in Physics–Electronics in 1992 at the University of Granada, Spain and a PhD degree in Industrial Engineering in 1999 at the University of Cádiz, Spain. He has four recognitions in the field of Communication Engineering, Computation and Electronics by the Spanish Government and was also awarded a Knowledge Transfer Recognition by the Spanish Government. Furthermore, he is a Full Professor in Electronics and founder of the Research Group in Computational Instrumentation and Industrial Electronics (PAIDI-TIC-168). His research interests include HOS, power quality and the inclusion of computational intelligence in technologies for measurement systems.

Dr José-María Sierra-Fenández received a PhD degree in Industrial Engineering in 1998 at the University of Cádiz, Cádiz, Spain. He is a Lecturer at the same University in the Research Group in Computational Instrumentation and Industrial Electronics (PAIDI-TIC-168). His research interests include energy technologies to manage energy efficiency and renewable energies, SG, power quality, instrumentation technologies, smart metering and HOS.

Manuel-Jesús Espinosa-Gavira received an MS degree in 2018 at the University of Cádiz, Cádiz, Spain. He is now a PhD student in Energy and Sustainable Engineering at the University of Cádiz from 2017 and is a Member of the Research Group in Computational Instrumentation and Industrial Electronics (PAIDI-TIC-168). His research interests include power quality, time-series analysis, sensor networks, meteorology applied to renewable energies and energy efficiency.

Dr Agustín Agüera-Pérez received an MS degree in Physics in 2004 at the University of Seville, Seville, Spain and a PhD degree in Industrial Engineering in 2013 at the University of Cádiz, Cádiz, Spain. He is now a Lecturer of Electronics in the Department of Automation Engineering, Electronics, Architecture and Computer Networks at the University of Cádiz and a Researcher in the Research Group in Computational Instrumentation and Industrial Electronics (PAIDI-TIC-168). His research interests include energy meteorology, power quality and virtual instruments.

Dr José-Carlos Palomares-Salas received an MS degree in Industrial Engineering in 2008 and a PhD degree in Industrial Engineering in 2013 (*summa cum laude*), both at the University of Cádiz, Cádiz, Spain. Currently, he is an Associate Professor at the University of Cádiz, in the Department of Automation Engineering, Electronics, Architecture, and Computer Networks and also a Member of the Research Group in Computational Instrumentation and Industrial Electronics (PAIDI-TIC-168). His research interests include power quality, intelligent systems and machine learning.

Acknowledgements

The work has been carried out in the framework of the following competitive Spanish National Research Projects:

TEC2016-77632-C3-3-R-CO – Control and Management of Isolatable NanoGrids: Smart Instruments for Solar Forecasting and Energy Monitoring (COMING-SISEM).

PID2019-108953RB-C21 – Estrategias de producción conjunta para plantas fotovoltaicas: Datos operacionales energéticos y meteorológicos para sistemas fotovoltaicos (SAGPV-EMOD).

In both projects, new techniques for power quality monitoring in the smart grid frame have been developed in the framework of the PAIDI-ICT-168 Research Group on Computational Instrumentation and Industrial Electronics (ICEI), founded by the Junta de Andalucía government.

During this research, a National Patent directly aligned with the method was proposed. ES2711204 Procedimiento y Sistema de Análisis de Calidad de la Energía e Índice de Claidad 2S2PQ, Caracterización de la Señal en un Punto Del Suministro Eléctrico.

In addition, researchers of our unit have been doing different four-month research stays at Dresden University of Technology, at the Institute of Electrical and High Voltage Systems Engineering, under the supervision of Dr Jan Meyer and Dr Ana María Blanco within the Power Quality Research Group.

Acronyms

AMI	Advanced Measurement Infrastructure
CDF	Cumulative Density Function
CENELEC	European Committee for Electrotechnical Standardization
CIGRÉ	International Council of Large Electrical Networks
CIRED	International Congress of Electrical Distribution Networks
dB	Decibel
DER	Distributed Energy Resources
DFT	Discrete Fourier Transform
DSOs	Distribution system operators
EV	Electric Vehicle
FDK	Frequency Domain Kurtosis
FFT	Fast Fourier Transform
HOS	Higher-Order Statistics
IEC	International Electrotechnical Commission
IEC-61000-4-30	Paper of the IEC
IEDs	Intelligent Electronic Devices
IEEE	Institute of Electrical and Electronics Engineers
LV	Low voltage
MV	Medium voltage parameters, such as RMS (root-mean-square)
PDF	Probability Density Function
PMD	Power Monitoring Device (device whose main function is metering and monitoring electrical parameters)
PQ	Power Quality
PQD	Power Quality Events Detection
PQ index	Proposed PQ index based on HOS

PQI	Power Quality Instrument (instrument whose main function is to measure, monitor and/or as certain PQ parameters in power supply systems, and whose measuring methods (class A or class S) are defined in the standards
PQIDif	Standardized PQ data format adopted to make data easily compatible between devices and establish data analysis procedures
Prosumer	A consumer capable to produce energy and consume energy
PMU	Phasor Measurement Units
PV	Photovoltaic Panels
RE	Renewable Energy
RTUs	Remote Terminal Units
RVC	Rapid Voltage Change
SG	Smart Grid
SK	Spectral Kurtosis
SNR	Signal-to-Noise Ratio
STFT	Short Time Fourier Transform
SWM	Sliding Window Method
THD	Total Harmonic Distortion
TSOs	Transmission system operators
U_c	Nominal Supply Voltage
U_{din}	Supply voltage variations
U_{ref}	Reference voltage
U_{rms}	RMS of the nominal voltage
V&I	Voltage and Current waveforms
WT	Wavelet Transform

1

Power Quality Monitoring and Higher-Order Statistics

State of the Art

1.1 Introduction

The importance of power quality (PQ) monitoring is that it becomes useful when detecting variable deviations and helps to describe the network behaviour and find solutions to some problems related to a wide range of electromagnetic phenomena concerning the interaction of power systems and end-user devices. The development of advanced monitoring solutions is essential to help utilities and end-users to understand network behaviour in order to discriminate, under a commercial PQ contract, the pollution source between the utility and the user.

1.2 Background on Power Quality

The principle of electrotechnical systems is to carry energy from the active element, the power source, to the passive elements, associated to the electricity consumers. Energy is then transformed into another type or simply produces work, through a conversion process that depends on the nature of the technological equipment used. During this process, the behaviour of the electrical variables involved, voltages (V) and currents (I), are mutually interrelated and the system behaviour is conditioned optimal performance (see Figure 1.1).

The PQ concept was introduced to establish the requirements on which the electric voltage would be delivered to consumers' terminals. Restrictions

Power Quality Measurement and Analysis Using Higher-Order Statistics: Understanding HOS Contribution on the Smart(er) Grid, First Edition. Olivia Florencias-Oliveros, Juan-José González-de-la-Rosa, José-María Sierra-Fernández, Manuel-Jesús Espinosa-Gavira, Agustín Agüera-Pérez, and José-Carlos Palomares-Salas.
© 2023 John Wiley & Sons Ltd. Published 2023 by John Wiley & Sons Ltd.

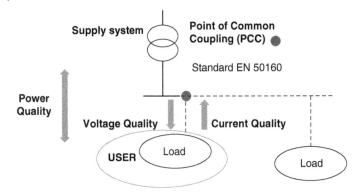

Figure 1.1 The EN50160 rules on voltage quality until the point of common coupling (PCC). *Source:* Authors.

were introduced on the method that would be used to supply power to consumers using the same connection point. Both conditions need to be accomplished on any network [1].

According to the standard specifications, supply voltages must not distort the load behaviour (quality of the supplied voltage) and consumers must comply with certain restrictions and not distort the voltage, inject harmonics in the distribution nodes or produce perturbations (quality of the end-users). Non-linear loads deteriorate current waveforms in a short period of time and distorted current waveforms (non-sinusoidal) increase voltage deviation. Finally, this kind of problem could impact the voltage behaviour in shared nodes and extend the problems to other loads.

The PQ concept is still evolving, a result of the gradual increase in the system's complexity over successive decades. Even though different monitoring solutions using higher-order statistics (HOS) have previously been proposed, there has been a need to elaborate on a compendium with a more holistic point of view that is applicable to future networks studies.

PQ phenomena consist mainly in detecting an electrical disturbance that is reflected in the power system as a waveform deviation when a sinusoidal oscillation increases and decreases their amplitude, symmetry, and waveform shape changes, mainly caused by circuit switching and the whole non-linear devices acting with unknown dynamics. Such network behaviour could be characterised using HOS. At present there is a lack of time-domain information in many studies because the complex solution that represents the data compression and storage challenges makes it difficult to correlate different data registers and indices. Most of the indicators that are extracted and stored come from frequency-domain information, which has a more complex representation format than time-domain information. In order to accomplish feature extraction, HOS can compress time-domain data and extract waveform characteristics. According to previous experience, the HOS domain seems a convenient strategy to use in order to compress data information while avoiding the highest computation complexity.

The present work focuses on the feature extraction of waveform characteristics using the sliding window method (SWM), which proposes the use of signal cycle-to-cycle processes from the time domain. The statistical information is translated in HOS planes, where the probabilities of finding different behaviour areas regarding different network states are greater than those employed by widely established signal-processing methods.

The book carries out a completely different proposal from that achieved by conventional PQ studies and is a step towards using the HOS application in the field. The most relevant topics to remark on the present work are:

- It characterises the voltage waveform distribution under normal operating conditions, defining the fundamental parameters and the limits from which a PQ problem begins (i.e. amplitude, fundamental frequency, symmetry and waveform shape). This last issue, has not being addressed before on an HOS analysis for PQ event detection or measurements in the field. This piece of information is relevant in order to conduct an HOS analysis closer to reality and to discriminate between the types of disturbances, depending on the time interval that considered during data analysis.
- It establishes continuous monitoring methods based on HOS. The use of feature extraction efficiently helps to improve significantly a more feasible network state recognition and increases the contribution of these techniques in advance to instrumentation for PQ management in future smart grids (SGs).
- Big data generated by continuous monitoring needs intelligent data management from HOS feature extraction and compression. In this thesis, data compression techniques are proposed both in time and in space.

The first chapter is an introduction to the quality of the electricity supply, as well as fundamental aspects related to its monitoring, signal processing techniques, efficient data management, new studies and the new infrastructure in the provision of network intelligence. A contribution of HOS as PQ indices in the advance-monitoring infrastructure in the context of the future network is also introduced.

The second chapter, centred on materials and methods for feature extraction from time domain analysis, summarises the parameters studied in the HOS domain. First, the standard and indices in the time domain that characterise the waveform shape are introduced. Later, by means of measurement solutions based on HOS, the waveform characterisation is presented. Finally, the different graphs and representations used in the subsequent sections are introduced to help the reader understand the different monitoring solutions.

Chapter 3 proposes methods that can be used to detect events based on HOS, extracting the different ranges for the subsequent analysis.

The fourth chapter introduces HOS measurements in the frequency domain, and although detected characteristics are not related to the wave, they relate to each of the frequencies that compose it. The indices that characterise the signal in the frequency domain are introduced and the HOS parameters studied as the spectral kurtosis are summarised as well as their ability to detect different harmonics.

Chapter 5 proposes monitoring a solution to find a PQ deviation index based on HOS. The experiment description is presented and the results show the contribution of each of the different strategies for virtual instruments using HOS in measurement campaigns.

As a result, a number of publications and a patent based on HOS for the characterisation of voltage signals in the power system have been introduced. The patented method helps monitor the network in order to inform not only suppliers but also network users about the problems of PQ and how the supplied energy is far from perfect. This establishes the groundwork that can be used to develop the advanced monitoring infrastructure, that is, future PQ analysers [2].

Figure 1.2 shows the timeline evolution of PQ monitoring based on the technology advances and users or actors involved in the different steps. Since the 1970s, domestic electrical power measurements have been taken using traditional basic concepts to satisfy mainly the production of bills and taxes to support the contractual relationship from producers, suppliers, and clients. The balance between produced and demanded energy has resulted in effective readings and are sufficient for this purpose. Under normal conditions at a commercial level, a higher rigour in the power measures is needed to manage the contractual cost or to control the installation.

Nevertheless, the use of normalised variables and network protection from external and internal disturbances have helped to guarantee and avoid electrical grid performance deterioration and reduce the impact during system operations or special events. Such situations usually impact consumers and the outgoings and fails on the service [5].

The massive introduction of electronic technology in residential and industrial installations has caused harmonic injection, transitory resonances in current and voltage, voltage surges, sudden drops or local voltage reductions, voltage and phase current imbalance in three-phase systems and many other instances, such as a decreasing power factor and phenomena of random manifestations often unknown or from blurry origins. This affects the

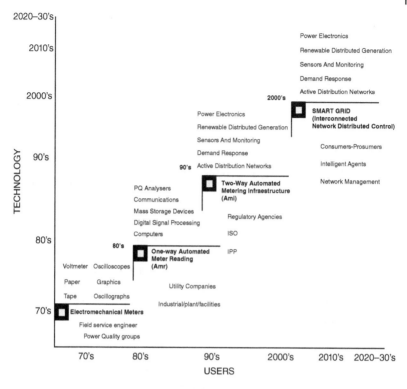

Figure 1.2 Time-line of power quality (PQ) monitoring technology evolution. The range of the first 40 years of the diagram was extracted from reference [3]. The smart grid (SG) approach comes from reference [4]. *Source:* Authors.

stable behaviour of the systems, causes failures and influences the electrical service security, so special attention is required. Furthermore, electric distribution systems are supported by large wired networks to cover wide areas, and each year this service is extended to new territories and cities, which keeps the electric market growing and staying competitive.

During the last decades, resources have been dedicated to install remote terminal units (RTUs), which are systems installed in remote nodes to monitor the electrical network behaviour close to consumer locations. RTUs allow control of the main system variables, collect the historical information about the network, identify and characterise the standard voltage profile near the client's installations, detect incidences, model and study the causes of perturbations and possible solutions, etc.

Nevertheless, one handicap is what to do with the amount of historical data generated?

Under regular conditions, most of the power and energy measurement equipment developed (analogue or digital) have been designed to function properly under sinusoidal signals, considering pure symmetrical and balanced three-phase systems. It is accepted that the methods and traditional measuring instruments for active, reactive energy, the power factor and the like will not produce realistic values in all scenarios or under real conditions. Researchers have pointed out some inconsistencies in the formulations used by algorithms of some of the most representative commercial digital meters used by electrical companies [6–8] in three-phase systems under non-sinusoidal conditions.

As an alternative, new measurement systems have been required. Since the 1990s, advances in applied signal processing and digital modelling have made it possible to introduce effective and robust algorithms to overcome such traditional limitations, reduce data storage requirements, increase measurement accuracy and guarantee the massive processing of measures over a wide digital measurement system.

The electrical metrology has introduced concepts and procedures to measure, study, classify and mitigate such phenomena as a last alternative [9]. These formulations have been corrected by the IEEE-1459TM-2010 standard [10] and some other approaches to the definitions have been proposed in reference [11]. These implementations have been possible by providing tools for the acquisition, processing and storage of real-time data in existing and new facilities, as well as by additionally allowing reading and automatic testing of meters, even at user level with hand-held PQ analysers.

New analytical technologies for automatic classification, such as pattern recognition, data mining, decision-making and networking, have appeared in the 2000s, helping to solve complex problems as part of the advanced measurement infrastructure (AMI) [3].

Nevertheless, when measuring in the field, it can be seen that the assumption of steady-state conditions in power systems is an approximation more related to the ideal conception of the network behaviour than to the PQ frame [5]. Under steady-state behaviour, two conditions of PQ conception must be met: supply voltage must not affect the load operation and the load must comply with the requirements of the network without distorting the grid or other connected loads. Nevertheless, under real situations, all the waveform parameters – amplitude, frequency, unbalance and so on – vary from time to time more or less based on dynamic conditions in the network [12].

In Europe, the Committee for Electrotechnical Standardisation (CENELEC) produced the standard EN 50160, which establishes the 'Characteristics of the voltage supplied by the general distribution networks' and defines the main characteristics as the voltage supplied by a general distribution network in low and medium voltages must be, under normal operating conditions, at the point of delivery to the customer [1].

Traditional PQ studies classify the different disturbances of the voltage signal according to the disturbance characteristics defined by the standard in advance, thus establishing a difference between continuous phenomena and events and capturing these different states in short time-series. In any case, PQ studies have been made whenever a previous problem has been detected in the network.

Nowadays, compliance monitoring of the electricity supply is not widely introduced in low-voltage systems for economic reasons, given the necessary deployment and the absence of an adequate infrastructure. In addition, there is a diversity of measurement methods relating to monitoring objectives or disturbance detection according to their duration. For example, there is a difference between the EN 50160 and the IEEE Recommended Practices for Power Quality Monitoring as both documents differ in their nature, as the first is a standard and the second a compilation of recommendations. In addition, they classify events differently.

The EN 50160 standard is more global and less restrictive in terms of network behaviour. Instead, the recommended practices have a lower accuracy with respect to the measurement time intervals and the magnitude of the events. For instance, this can be observed where there are long interruptions; such cases can be computed in more detail by the IEEE Recommended Practices, to accuracies above or below one minute.

As to instrumentation, it has already been detailed how hardware and software technologies have been evolving and improving to give a better performance of measurement chains and reduction of computation effort of microcontrollers. The continuous measurement systems include sensors and intelligent electronic devices (IED) that manage acquisition and carry out evaluation analyses.

PQ measurement methods are ruled by the standard UNE-EN 61000-4-30 [1]. Focusing on the measurement unit in PQ devices, all input signals – voltage and current – are computed and processed by extracting different indices. As the network evolves, the indices used in feature extraction must also evolve. Actual indices are not new; they have already been used for several decades. The main references are IEEE 1159 [12] and UNE-EN 61000-4-30 [1].

Measurement indices that characterise different PQ disturbances are needed for different objectives [5, 12], such as:

– Accurate measurements under contractual applications, in order to verify the agreement of the contract constraints and detect where the network is disturbed.
– Compliance test helps to verify the compliance in terms of emission and immunity of different devices to meet the network conditions provided by the standards.

- Troubleshooting, which focus on measurements to detect the origin of malfunctions in the installations caused by different devices and systems that trigger protective equipment. Monitoring troubleshooting helps solutions to be found or corrective actions to be taken.
- Statistical surveys based on measurement campaigns in the power systems help to identify the level of PQ according to the regulations and standardisation.

The measurement purposes of the application determine the accuracy of measurement and the devices involved in the monitoring campaign. Figure 1.3 shows the measurement chain proposed by UNE-EN 61000-4-30. In the case of compliance verification, mobile or fixed power quality instruments (PQIs) are recommended according to UNE-EN 61000-4-30, class A [1]. In statistical surveys, a fixed PQI is according to class S [1]. Power monitoring devices (PMDs) are also usually suitable, leaving mobile PQI for cases of limited measurement duration. Site characterisation studies place mobile PQI according to class S [1]. In the case of troubleshooting, the best option is mobile PQI because it has high flexibility and analysis functions. Compliance with the standards is not necessarily required.

Additionally, when monitoring the network, digital measurements could be impacted by a non-adequate equipment selection and the acquisition of voltage and current data. There is a greater risk of uncertainty than those specified in the class of permissible devices [13]. Therefore, it is necessary to consider an adequate balance of errors between measurement methods and device extension when it comes to these complex techniques.

Nevertheless, whatever the application is to any monitoring campaign, it should demonstrate the capability of providing reliable information about the problem to be monitored. Additionally, monitoring devices must have the precision required for the measurements to be performed, and the methods and the monitoring campaign must be carried out in an efficient and economical way [5].

In practice, until this last decade, PQ studies have become special studies dedicated to characterise a number of specific applications in the industry,

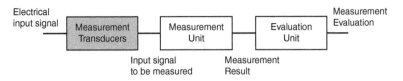

Figure 1.3 Power quality (PQ) measurement chain proposed by the standard UNE-EN 61000-4-30 [1]. *Source:* Authors.

such as wind turbine connections, harmonic detection, phase imbalance, the connection of large energy blocks or tracking specific clients in medium voltage (MV) to low voltage (LV).

Nowadays, unless there is some malfunction indicated in the network, PQ is not measured at the connection points of end users. Figure 1.3 shows the PQ measurement chain proposed by the standard UNE-EN 61000-4-30 [1].

1.3 PQ Practices at the Industrial Level

Existing practices at the industrial level are summarised in reference [14], which is a survey of 114 questions made in 2012 to different transmission system operators (TSOs) and distribution system operators (DSOs) from 43 countries, with a presence in all continents. Results show that there is no preference on the type of phenomenon to be detected at that level. The monitoring objective is to obtain a global state of PQ behaviour instead of just a survey of individual PQ phenomena. From the stakeholder's point of view, DSOs are more interested in detecting the majority of PQ events than TSOs. Both prefer to monitor line-to-neutral voltage measurements.

Most of the surveyed companies (86%) use the software packages attached to PQ monitors. In this sense, the instrumentation industry plays a crucial role in terms of data management and the necessary quality in terms of compatibility, support of different formats, software requirements, etc.

As to the use of PQ equipment, it is remarkable that only a few companies control a large part of the market, as it is a favourable factor in terms of standardising the handling and processing of PQ data.

On the other hand, there is a great diversity of methods for event-based detection and management based on time-scales [15]. The issuance of reports is more frequent as long as it deals with certain events. However, 32% of the surveyed system operators prepare reports based on specific time periods – monthly or weekly. Other reasons for generating reports are factors that affect a certain consumer systematically or after an event. However, such a report does not include short-term measurements to verify customer complaints or solve external problems.

1.4 A New PQ Monitoring Framework

Taking all these premises into account, the electricity market has continued to grow and today it follows a strategy of simultaneously evaluating new technologies while developing tools to face new types of technological problems, such as big data management, which comes from multiple application tools [16].

Moreover, the power system today is evolving through a process of multiple changes and challenges towards a more flexible power network or SG [17]. The main issues relating to that transformation are the decarbonisation process, decentralisation, the introduction of renewable energy (RE) and storage, distributed energy resources (DERs) and the energy management through the introduction of information and communication technologies.

In the last five years, there has been a growing number of publications and books dedicated to the SG in power systems (Figure 1.4). Compared to the SG, PQ is a much older field that has achieved a high level of expertise and even had a slow trend to decrease in the last few years (Figure 1.5). However, it is warned that the PQ field will introduce drastic changes in its conception with the opening of new research lines in the future network [18]. In this scenario, PQ is an important issue of the power system and cannot be neglected. The future PQ research will play a relevant role in the transition to a more flexible network (Figure 1.6).

Researchers have been discussing the need to propose another approach in order to solve some future challenges related to this issue. The main factors related to PQ in future grids would be the new generation equipment, end-user devices and the changes within the grid coming with the new solutions and their interactions. Achieving an adequate level of PQ guarantees power system compatibility between consumer equipment and the grid operations itself.

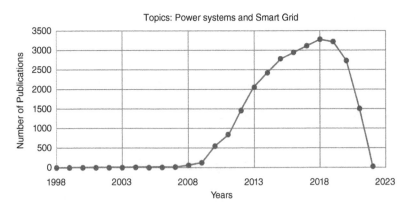

Figure 1.4 Evolution of publications related to power systems and smart grid (SG). *Source:* Web of Science (WOS).

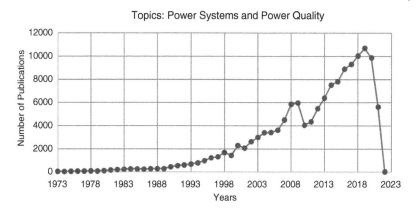

Figure 1.5 Evolution of publications related to power systems and power quality (PQ). *Source:* Web of Science (WOS).

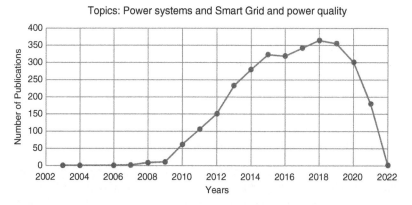

Figure 1.6 Evolution of publications related to power systems, smart grid (SG) and power quality (PQ). *Source:* Web of Science (WOS).

In the next section, the state of the art of the PQ in the era of the SG is introduced, as well as the new measurement and instrumentation methodologies brought about by the solutions proposed in this book.

1.4.1 The Smart Grid

The smart grid (SG) is a concept for electricity networks across Europe. The initiative responds to the rising challenges and opportunities that bring benefits to all users, stakeholders and companies that perform efficiently and effectively [19].

Focusing on the SG concept, there are multiple definitions. However, the task force on the SG was proposed by the European Commission, which uses as a definition of the specific European challenge: 'Future grid that is needed for reaching efficiently the EU Energy and Climate Change targets for the year 2020'.

Some other definitions put the focus on the use of the technologies to solve the main challenges of the grid. According to the International Electrotechnical Commission (IEC), a definition for the SG would be the electric power system that incorporates information control strategies, distributed computing and associated sensors and actuators for purposes such as:

1) To integrate the behaviour and actions of the network users and other stakeholders.
2) To efficiently deliver suitable, economic and secure electricity supplies.

Furthermore, the Institute of Electrical and Electronics Engineers (IEEE), propose the next definition on their smart grid website: https://smartgrid.ieee.org/about-ieee-smart-grid.

The smart grid has come to describe a next-generation electrical power system that is typified by the increased use of communications and information technology in the generation, delivery and consumption of electrical energy.

1.4.2 The Smart Grid and the Power Quality

As previously introduced, from many decades and until today, one of the basic aspects of system performance remains and improves PQ levels on networks (Figure 1.2). The SG concept improves network flexibility with 'smarter' technologies that provide functionalities, operational principles and services with a higher level of security. The sustainability and affordability of the electricity supply will help to achieve higher PQ levels on future networks. Additionally, new electrical tariffs and consumers' and prosumers' empowerment is expected.

The Smart Grid Reference Architecture Model (SGAM) proposed by the CEN/CENELEC/ETSI Smart Grid Coordination Group defines a three-dimensional model that includes five intraoperative layers: component layer, communication layer, information layer, function layer and business layer [2].

According to such a model, the future grid will increase its complexity with higher numbers of renewable-based variables along different domains and zones. The supply demand interactions will change throughout the traditional ranges with new loading conditions, demand responses and manage resources of technologies that will be introduced. Supply-side solutions, as conceived now, will not be enough to respond to the new dynamics of two-way power flows and the uncertainty that comes with the introduction of renewable sources. Contributions are expected to come from coordinated demand side control in adaptable and flexible power.

The monitoring solutions between the service provider and customers will take place on the information layer, which 'describes the information and data models that represent the functions or services solved through the communication'. Indeed, data acquisition and device management solutions are going to deal with a high efficiency, intelligent and flexible control within the information layer such as that described by the Smart Grid CENELEC Group. Power systems are operated differently to traditional power system utilities in a more flexible way. An efficient management of all these issues will ensure that their PQ will not deteriorate and will deal with the challenges between the grid and the connected loads. The increased sensitivity of loads connected to the grid will involve end-user interactions as a result of the power electronic devices and their interference with power lines and other communication systems in the grid.

In this deregulated scenario, there are several trends associated with PQ monitoring, each of which demands specific solutions and involves multi-disciplinary actions. New tools and monitoring solutions, such as snap shots, are required that will help describe the transition from past systems towards the future scenarios considering PQ issues and the high cost of measuring each node of a large distribution system [5].

In this sense, there is a concern in the scientific community about new PQ discrete and non-discrete disturbances. Some implications might be the introduction of higher harmonic emissions, switching transients more often, voltage sags and swells, voltage short interruptions, DC offset and short unbalances [20]. All these PQ issues will change the probability of interference in the SG and there is a high potential occurrence of new types and multiple disturbances with the introduction of different technologies and renewable resources. Some advances in power electronic controllers will increase the network harmonic emissions and other new events. In this sense, advanced voltage control from multiple locations will be needed. In addition, renewable resources such as solar panels produce surges, wind turbines can cause high frequency signals from the switching frequency converters, electric vehicle (EV) chargers emit harmonics and heat pumps are sensitive to flicker when connected. To date, these issues are far from solved and need to be the focus of future PQ studies, aligned to monitoring solutions capable of helping to characterise the new scenario [20].

1.4.3 Performance Indicators

In contrast, what really matters to actual end-customers are mainly the energy price based on the network tariffs, the reliability of the system, supply continuity and PQs. Alternatively, some special customers will continue to be concerned about the environmental and safety issues and will increase their interest in the energy field.

In addition, the technical aspects of PQ will remain under discussion within the scientific community and new ways to adapt their contents will gradually be adopted. However, power system stability and their coordination are technical issues about which customers will not be involved. PQ knowledge is important in order to obtain suitable and reliable indices and limits to study the hosting capacity impact of the power system.

Future SG must be reliable and must solve some challenges that have already been mentioned. Thus, with the introduction of DER, the concept of hosting capacity of the system plays an important role [19, 20]. From a general point of view, the hosting capacity are the limits from which the system 'hosts' DER, such as distributed solar, energy storage and (EVs), on a feeder. Acceptable reliable energy limits must be designed in order to accomplish PQ to the customers. Already existing and new mitigation methods will ensure the electromagnetic compatibility (EMC) in future electricity networks [21].

PQ affects technical and economical decisions in the power system, when connecting more distributed generation than the hosting capacity of the system can hold. If nowadays standards, measurement methods and requirements do not adapt to future challenges, they will remain a limitation to the advance in the PQ field. In this sense, it is not only relevant to measure the actual standards that are demanded, but also to have a new vision in monitoring strategies that are needed.

1.4.4 Existing Measurement and Instrumentation Solutions

There are numerous PQ measurement analysis procedures depending on the network point, the monitoring objectives of the study and the customer needs.

The new instrumentation will perform voltage and current measurements with an accuracy that is not presently accomplished by PQ studies. From now on, PQ measurements will depend on the studied phenomena based on network configuration and monitoring objectives.

Some main aspects related to the instrumentation for PQ shown in Figure 1.3 are summarised in the following way:

1) PQ monitoring location.
2) Hardware solutions for PQ monitoring, including transducers, sensors, IED accomplishing the network requirements and a communication interface.

3) Measurement units, including measurement indices.
4) Analysis of measured data from an evaluation unit.

Nowadays, PQ measurement locations are solved by utilities promoting monitoring substations and customer PCC. Indeed, the advanced measurement infrastructure (AMI), such as protection relays and controllers, are deployed in multiple networks at this level. However, notice that the majority of these devices do not exhibit storage capabilities although this is the potential infrastructure to integrate within the SG.

Additionally, smart meters were deployed throughout Europe, but most of them do not incorporate PQ monitoring solutions. The focus of what to measure in each location is based on the customer needs, data storage and new ways to manage their data. Until now, PQ in LV is not measured with a traceability level comparable to the regulations, only when it is a special customer as, for example, in a factory or when a monitoring campaign is taking place with fixed or hand-held devices [12]. On the future grid, monitoring in customer installations, based on their local thresholds and network behaviour, would help with many of the challenges that distribution feeders currently face.

With regards to hardware capabilities, IED solutions that process signals, transducers with voltage, current, temperature and other parameters are needed. According to international tests performed on calibration laboratories, transducer manufacturers must provide more information related to the frequency response of their devices and the calibration processes for magnitude, phase angle and temperature correction [22].

Related to the IED, during the last few decades the quality of the supply was monitored using PQ analysers, scopes and other infrastructures used to measure the PQ delivery by utilities (DSO and TSO) [23]. Nowadays, utilities are experimenting with IED capabilities in data acquisition, analysis, storage and transfer to the control centre. A combination of IEDs, such as a controller, meters/AMI and relays, are capable of monitoring line currents, line to ground voltages, capacitor current, neutral current, etc. [24]. Some of them comply with IEC 61000-4-30 in measuring harmonics, voltage unbalance, dip, swell, interruptions, inter-harmonics, etc.

The measurement unit of PQ instrumentation adopts the standards measurement methods and indices. Nevertheless, the instrumentation society and researchers in the PQ field have been arguing about the need to adapt many of these techniques to the new challenges. Following the SG, devices connected to the grid have a communications interface, a degree of intelligence based on their ability to process data of signals and share the results with a level of control [25]. In the SG context, many DERs, which can introduce new events that impact PQ, are foreseen. They are not detected by conventional instrumentation since this includes its design based on previously well-known disturbances and some new ones are not characterised enough [4].

Future networks will offer more flexible architecture configurations and some other capabilities, such as connected or islanding operations as on microgrids. Nevertheless, none of these capabilities will be accomplished correctly if another capability is not solved, which is the continuous measurements and monitoring of the fundamental electric parameters, such as voltage, current and other environmental variables (e.g. temperature, humidity, water level or gas presence).

In the next sections, the instrumentation challenges and signal processing techniques are summarised in order to provide a complete overview.

1.4.5 New Approach in Measurement and Instrumentation Solutions in the SG

Until now, the research community has focused mainly on both approaches: the evaluation of standardised indices and monitoring the network with individual indices based on well-known event detection. Therefore, in a future scenario the focus must be on monitoring in the field. Future instruments need to be more flexible and adaptable to new measurement campaign approaches.

A new trend in monitoring is the proposal of non-disturbance state-based processing vs. the traditional disturbance objective approach. While traditional PQ analysis has been devoted to detect and storage, a pre-conceived disturbance is based on their associated index. In the future, the end-user will be provided with a greater flexibility and future instrumentation will implement detection of different states solutions. Moreover, some researchers have introduced deviation indices that measure the deviation (i.e. distance) from an ideal state when different disturbances appear [26].

Feature extraction in future signal processing will not be limited only by traditional indices. The aim is to propose a new analysis and methodologies to process data that can be compiled and compared with that of the pre-processed data. New ways to analyse and process data will be involved in statistical analysis, which will help in the network transition in evaluating data, characterising the network and validating the new and demanded indices into new PQ monitors. Indeed, new indices are proposed in the literature to characterise each of the disturbances [4]. Future practices would not exclude current practices and so they may include them based on PQ monitoring objectives and the expected impact of a future network, such as the expected impact of the future methods for volt-var-control in the distribution grid [27].

Additionally, continuous monitoring is the best strategy to process and detect all kinds of waveform distortion phenomena, even more so when the different standards discriminate between events based mainly on their duration. In addition, the development of more comprehensive indices helps to describe the network changes through the statistic extracted from waveform characteristics.

To simplify reporting using a lower number of indices is desirable, and is still under discussion today in the researcher community. The status of the system needs to be visualised and easy to interpret regarding the flexibility to report PQ on future networks with indices and data visualisation solutions in advance [28, 29]. To improve PQ visualisation helps when evolving or mitigating the impact of such PQ phenomena, avoiding the need for others to increase [30]. Furthermore, new challenges are expected on PQ indices characterisation through the introduction of DER. For example, new indices are aimed to be introduced to understand the unbalance caused by the shading in photovoltaic (PV) power plants [21].

1.4.6 Economic Issues for PQ

Nowadays, the cost of PQ has been the cost of a failure or a malfunction of a device. On future networks, in which the consumption and energy production will be distributed, small consumers or micro-network managers will influence the balance of the system [31].

Some technologies that will help in the transition to the SG are consumer electronics, PV panels, EVs, energy-efficient lightings, electric heat pumps and others that will play an important role incorporating the PQ issue. Technologies based on the storage of energy in batteries, microgrids and virtual power plants will help to mitigate the impact of voltage variations, although it is true that they demand an initial investment. In addition, with the deployment of smart meters in Europe, several technologies have been introduced, some of which include PQ functionalities. However, the PQ criterion in metering still needs to be homogenised [32].

In addition, IEC 61000-4-30 includes a section that establishes the bases for contracting energy quality. In general, the new equipment must include additional hardware costs with the objective of introducing PQ reports. Nevertheless, no reference solutions to instruments data management and visualisation of the big data are included in any of the instrument classes introduced in the standard. Therefore, it is not only necessary to quantify the levels of damage on the network and the origin of the failure but also to establish responsibilities between customers and producers [33].

It has been warned that there will be a cost distribution for the PQ issues among all stakeholders in a proportion of economies of scale. Network operators will have to bear the initial costs of high network distortions but, subsequently, through regulatory schemes, they will be likely to recover these through tariffs. Moreover, manufacturers can see their costs increased if they are called to perform more tests on the products. The higher probability of interference will result in higher costs for users in which their network is insufficiently immune, especially against voltage variations and harmonic emissions due to the introduction of more power electronic converters connected to the grid [20].

All in all, the market must consider the cost of keeping the PQ at its limits. Cost distribution through the stakeholders will depend on the market structure and will influence the future regulation of the PQ [34].

1.4.7 Power Quality and Big Data

The deployment of smart meters in Europe has introduced the concept of metering that is expected to become decisive in the field of PQ with the introduction of RE. As discussed in the previous sections, monitoring devices collect data according to traditional standards, based on specific thresholds and aligned to detect certain events. Still today, this is a barrier to development of the advanced measurement infrastructure (AMI), monitoring equipment and applications on a continuous basis.

In addition, algorithm development needed to be implemented in a PQ infrastructure is an important contribution in PQ knowledge. It is necessary to solve the gap between data management and automatic methods.

In a few years, the big question will be how to manage all data that comes from continuous monitoring and storage. Even when storage and data processing challenges are difficult, the most essential knowledge is finding useful and valuable information through signal processing and feature extraction of the network status. Signal compression techniques (i.e. V, I) are not as mature as image and video compression [25]. The next section is devoted to different techniques.

Any monitoring system needs a balance between processed information and its storage capabilities. Moreover, getting information without storing all the information using thresholds and searching the most suitable issue each time demands an advanced knowledge about the probabilities of the analysed population.

Additionally, the PQIDif format [35] allows large amounts of data to be standardised in terms of manageability, traceability of measurements and data reading from measuring instruments [36]. The standard is adopted to make data easily compatible between devices and establish data analysis procedures. Nevertheless, this standard does not improve the big data visualisation issues, which are more and more important in this context.

As a result, PQ monitoring objectives in the SG will spread and more information would have to be managed. Detailed studies most consider the impact of new equipment on the grid. Non-steady state chargers, such as ovens, TV, the massive introduction of wind and solar power, and EV chargers, are not characterised enough. Indeed, changes in frequency below 49 Hz would impact the performance of the different loads. In that context, the fast voltage variations, below one second, a few seconds or a few minutes, would affect the network state. However, it cannot be ruled out that even relatively minor frequency variations might adversely impact the control algorithms for certain power electronic converters.

Figure 1.7 introduces the frame on which new challenges in the PQ field are studied, through deploying smart signal processing tools and indicators.

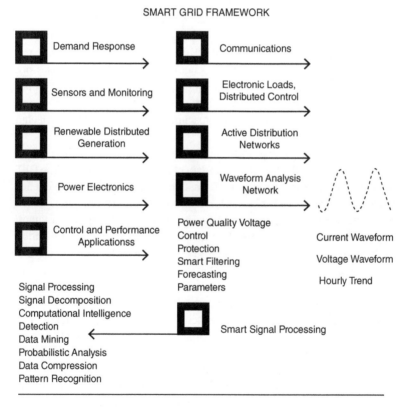

Figure 1.7 Power quality (PQ) measurements in the smart grid (SG) context extracted from CIRED/CIGRÉ Publications reports. *Source:* Authors.

1.4.8 Signal Processing for PQ

Existing PQ indices are related to both domains, time and frequency. Consequently, a summary of the signal processing techniques that are most used in detection and classification of events is made in this section. Signal processing analysis has three main different steps: pre-processing, feature extraction and automatic classification schemes.

First, the pre-processing stage demands parameter estimation based on signal segmentation under different states (i.e. before, during and after an event takes place). In addition, detecting the transition states is important as well. Second, feature extraction is based on different analysis techniques on time and frequency domains that emulate the most efficient waveform characterisation. There are many feature extraction solutions in an analysis of power systems. The simplest analysis in the time domain is made from the detection of the fundamental frequency amplitude estimation through root mean square (RMS). Some advantages of the RMS method are that it is possible to calculate cycle-to-cycle with a high speed of calculation and can easily detect voltage transients or sags. However, it is sensitive to noise and depends on the analysis of window length – one cycle. Also, it does not distinguish between fundamental frequency, harmonics or noise components. As phase angle information is lost, events such as transformer saturation, induction motor starting and capacitor switching cannot be detected using the RMS method.

On the frequency domain, the most common technique is the discrete Fourier transform (DFT), which is a variant of the Fourier transform. The DFT method transforms the waveform information through time to the frequency domain and detects the fundamental frequency amplitude and phase angle. The method performs characterisation on periodic signals in the stationary state. Conventional indices based on the Fourier transform have the limitation that they are not able to characterise non-stationary or aperiodic signals. An advantage of this method is their fast computation because they measure the fundamental amplitude and harmonics content. However, the window length depends on a minimum frequency resolution of 10 cycles.

Additionally, some other feature extraction techniques propose to improve DFT limitations. For example, filter banks and adaptive filtering both help when extracting signals from a specific bandwidth, low band, band-pass or high-pass band. One advantages of filtering the waveform is that they are not corrupted by noise, help to detect rapid changes and offer the possibility of a most compact implementation of single filters. Some other techniques have been derived from filters; an example of this might be the weighted least squares (WLS) method.

The Kalman filter is another signal-processing tool devoted to study the specific frequency band spectrum in power systems. It is defined as a state-space model and helps to monitor fundamental frequency and different harmonics amplitude and phase angle. It is immune to noise, although it demands a higher calculation complexity [37].

Since the middle of the 1990s, the wavelet transform (WT) has been introduced in power systems to perform non-stationary analyses of harmonics. The WT has been widely extended in PQ signal processing in order to solve some limitations of indices based on the Fourier transform, providing suitable time-frequency resolution. Nevertheless, wavelet-based techniques present a complexity in terms of computation, as well as tolerance to the signal noise and the difficulty to automatically classifying the detected disturbance. Wavelet analysis is highly dependent on what the mother wavelet chooses. The wavelet disadvantages are the main reason why there is not a universal use of the technique, a standardisation use and a consensus within the research community.

The short time Fourier transform (STFT), commonly known as the sliding window version of the DFT, comes from an extension or concept combination of both wavelet and Fourier transforms. It is a particularization of the associated mother wavelet and uses a Gaussian window in the analysis. The STFT is capable of measuring the signal, offering both information, time and frequency related to disturbances. Nevertheless, fast-changing high-frequency components (i.e. transient signals) cannot be computed widely following this method because of the fixed width window function incorporated by the STFT method.

On the other hand, it is easier to detect and classify events using the STFT rather than the WT. In addition, the STFT is also immune to noise and its implementation is based on an extension of the well-known DFT algorithm. When noise environments need to be analysed, techniques such as the STFT and the Kalman filter are suitable. However, these techniques exhibit complex calculations and high computational cost [38–40].

As the last step in the signal processing procedures, automatic classification studies involve classification and decision-making processes through different pattern recognition techniques and artificial intelligence, fuzzy logic, neural networks (NNs), and others. Some previously mentioned techniques propose to identify, classify and diagnose power systems evolving through automatic methods focused on specific objectives. For example, classification and analysis of voltage dips use the Kalman filter for the amplitude estimation.

Wavelets have been applied to automatic classification in many event disturbances such as harmonics and transients [41–43]. The algorithms have been applied in a single site or in specific points of the network. Once more, feature extraction based on wavelet techniques exhibit many disadvantages. In the case of fuzzy logic, the knowledge is based on a set of statistical rules in order to pre-process the time-series. They offer rapid computation with human line reasoning capabilities. Fuzzy logic methods have the ability to deal with imprecise information. Nevertheless, fuzzy classifiers need the opinion of heuristic experts to solve data relationships and they lack self-organising and self-turning analysis [37]. They have been combined with wavelet features of PQ events and fuzzy classifiers, but the recognition is reduced if the signal presents noise and the data set used to train the system is not real. The complexity of NNs is related to the training time and the error of divergence depends on the systems' complexity. The limitations of NN are that they are unable to manage imprecise information, solve conflicts and rely on a hidden layer trial error. Therefore, NN needs to be trained for specific purposes [37].

These automatic methods and some other optimization techniques are dependent on different network conditions, though most of them are previously trained to find specific events. Their calculation is complex because they combine different procedures. The expert systems are not trained enough under multiple events and would be overstrained with the risk of losing selectivity. In addition, it is difficult to adapt these expert systems to different networks and to the network operator's requirements. Nowadays, the application of these techniques is limited and none of their applications are standardised.

As previously mentioned, feature extraction is an essential step in accomplishing a successful signal processing analysis. Most of the signal processing literature in the PQ field focuses on classification schemes of the statistical system or tools. This dissertation chooses as its technique the higher-order statistics (HOS) domain with the objective to fill the gap between the best feature extraction and the steps in events classification. This information is needed to incorporate pattern recognition into PQ instruments with a holistic overview. A feature extraction overview based on HOS aligns with the advance technical instrumentation.

1.4.9 HOS for PQ Analysis

Gaussian signals are characterised by statistic variables such as the mean and the variance, so in this case HOS are useless. Nevertheless, in practice, many real waveforms exhibit non-zero mean HOS and many noise processes have a Gaussian origin. Researchers have demonstrated that noise processes are negligible to HOS computation [44]. HOS have been traditionally used in different statistical studies detecting the changes in the waveform distributions. It is in these distributions where the revealing aspects in the 'shape' of the waveforms appear. When discrete data are available but the distribution of the variables remains unknown, estimators based on HOS can be useful.

Therefore, HOS have been used in applications that demand a characterisation of non-linear, non-stationary and non-Gaussian nature signals. For example, these could be in the fields of structural vibrations [45], detection of mechanical vibrations [46], acoustic detection of pests such as termites [47], detection of non-linear behaviour and the feature extraction in signal analysis in the frequency domain through spectral kurtosis (SK).

More than a decade ago, the HOS were introduced in PQ feature extraction studies. Compared with other techniques such as DFT, STFT, WT, RMS or other time-frequency signal analysis and second-order methods, they are less sensitive to noise processes and can help detect changes of 2% in the nominal voltage frequency that reduce the performance of the monitoring system [44, 48]. Additionally, they help to reduce the computation complexity in field measurements. Indeed, the spatial representation of HOS helps to detect and classify disturbances.

Figure 1.8 shows the evolution of HOS related to publications in the PQ field. In reference [49], the idea of using HOS has been centred on feature extraction in event detection and characterisation. HOS have been used mainly in applications of power quality disturbance (PQD) events detection. The main reason for using HOS in PQ studies is their capability to characterise non-Gaussian systems and processes like transients that occur in a power system. In addition, it has been shown that HOS are immune to Gaussian noise [49–52], which confirms the robustness of methods based on HOS under noise processes in power systems [53].

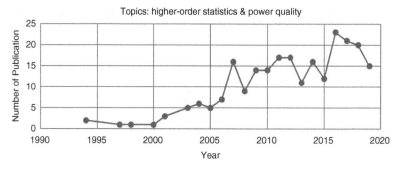

Figure 1.8 Evolution of publications related to higher-order statistics (HOS) in Power quality (PQ) studies. *Source:* Web of Science (WOS).

Several publications have used detection methods based on HOS in time and frequency. According to the calculation complexity, it is easier to calculate and extract the cycle-to-cycle statistics of a signal from the time domain. In the frequency domain, the SK combined with DWT are derived from HOS analysis in PQ solutions. This is a more complex analysis because it demands more complex calculations and representations. A discussion and interpretation on HOS in time and frequency domain were made in references [53] and [54]. Also, it is possible to recover some feature signals such as amplitude and phase.

Event classification through HOS has been done for individual kinds of events. Different characterisation studies have been done. Some have focused mainly on the detection of transitory disturbances [55]. Bi-dimensional planes of maximum and minimum cumulants were found in regions related to transient events.

In addition, sag and swells were also detected extracting HOS from the time-domain analysis, without considering changes in the fundamental frequency and the phase-angle jumps, at the beginning and at the end, when events had taken place [55]. However, the characterisation study shows HOS robustness under noise when measuring from SNB at 30 dB without being impacted. None of these studies of HOS feature extraction has demonstrated how the fundamental frequency changes impact HOS measurements.

Harmonics and inter-harmonics have been previously detected using HOS [56]. The predominant frequency on the probability density function (PDF) of the processed signal has been detected when extracting HOS for these purposes. In reference [57], inter-harmonic detection and identification have been made in order to identify a number of inter-harmonic components in a waveform according to different predefined classes. This information is helpful in order to improve spectrum interpretation related to inter-harmonic detection. In reference [58], sub-harmonic detection and identification using HOS was implemented. Depending on the nature of the deviation, a cycle may suffice to characterise a steady-state disturbance, which makes the methods based on HOS related to a characterisation on the frequency domain convenient for this issue.

An application to the SK for characterisation of PQ events [59] shows a WT-based witness procedure and offers a heavy data matrix difficult to implement in the traditional PQ instruments. In addition, reference [60] makes a characterisation of sags coupled with transients and harmonics at different noise levels. The feature extracting method based on SK is complex. Noise robustness is good, but the sag length is not detected by the SK.

Detecting the origin of events in characterisation of power systems will depend on the robustness of classification methods based on HOS. Most of the methods are devoted to classifying a limited number of disturbances. Some of them remain with an unknown probability of occurrence in the power system because this topic has not been studied enough through this perspective. Consequently, future methods will need to implement on-field measurements that are better aligned to customers' needs and network characteristics, which is where a more realistic framework probably lies.

Some of these solutions are even covered by the field of virtual instrumentation (VI), which offers a solution through a hardware support and measurements in an off-line detection based on previous captured events. In reference [61], a virtual instrument based on a LabVIEW™ platform that detects sag and swell and discriminates from other events was proposed. In addition, case-based reasoning helps to classify events among five categories. A HOS-based virtual instrument for PQ assessment [62] was made to show the computational capabilities of HOS when using real-time platforms.

In advanced classification studies, HOS were used as a feature extraction combined with different computational intelligent tools. For example, a categorisation of transients using HOS and competitive layers-based NNs [63] shows that maximum and minimum regions can be detected and discriminated in the HOS planes when a transient takes place. In reference [64], HOS (fourth-order central) and classification based on self-organising NNs helped to detect different regions in a data set of 90 signals with different anomalies. From the experience in reference [64], set foundations were found in automatic classifications based on HOS, but it was concluded that additional features were needed to face the problem of subclass division. Indeed, short and long duration transients have been classified into different regions, following a similar strategy based on HOS and NN.

In a recognition study of impulse and oscillation, transients were based on WT, the SK and the neural network. However, the number of training samples of RBF NNs affects the recognition rate [65]. As previously mentioned, NN demands a training stage and the weight in their efficient data characterisation is the robustness of the data set in capturing the monitoring objectives.

Oscillation and an impulsive transient can be recognised from characteristics extracted using incomplete S-transform and Morlet wavelet SK [66].

Most of the state of the art related to HOS in PQ are approaches that only have regarded synthetic signals as input in characterisation studies. Despite the advance in the PQ field, studies based on HOS have been more devoted to the robustness of computational intelligence methods to classify events than to interpret the parameters and the information related to electrical behaviour that is feasibly extracted through HOS. A problem with these parameters is that they are very sensitive to isolated extreme data since the values of the variables enter the cube and are fourth in its calculation. Because of this, it is important to improve their introduction to PQ studies.

A different approach using HOS gives importance to the detection of characteristics in the signal and their analysis has been combined with other indicators such as the Shannon entropy, a high/low power spectral density ratio and a maximum gradient [67]. However, they are separated applications; there is no continuity in these studies [68]. It is complex to assume these signal processing comparative techniques without having a background devoted to each one of them in the PQ field.

Although these previous studies only focus on the detection of events, it is fairly demonstrated that the extraction of statistical characteristics of the signal both in the time domain and the frequency domain are possible through HOS. The feature extraction time domain analysis seems to be the easiest technique, but it must face the dynamic characterisation of events in the time domain.

Hence, Figure 1.8 shows that HOS is not widely extended in the power systems signal processing. Nevertheless, in the PQ field, the HOS does not constitute an objective in the regulations related to PQ, even when feature extraction is easily compared with the previously mentioned traditional signal processing methods. Moreover, there are not many researchers specialised in the HOS field aligned to PQ monitoring campaigns.

Finally, continuous analysis is easier, extracting the information through the SWM. Through three statistics values, it is possible to gain information about the amplitude, symmetry and sinusoidal conditions cycle-to-cycle in order to simplify PQ monitoring in the SG and improve the advanced pattern recognition studies. This is one of the main challenges that this book makes to open the door.

2

HOS Measurements in the Time Domain

2.1   Introduction

This chapter discusses the main concepts related to higher-order statistics (HOS) methods that will be used in different measurements applications within this document. In addition, power quality (PQ) standards are specified.

In this section, the last advances on time-domain methods based on HOS are introduced. The concepts of scalability and compression over time are discussed and how they are related to HOS monitoring.

In Europe, PQ refers to the quality of the energy service and how it is administered and how it must follow the criteria established by the standard UNE-EN 50160 [13].

The continuity of the supply is relative to the number and duration of supply interruptions:

- The quality of the product, relative to the characteristics of the voltage waveform.
- The quality of the attention and the relationship with the client, relative to the informative actions and customer advice, hiring communication and claimed actions.

The power supply quality summarises the continuity, reliability and quality of this service. A few years ago, the main concern of electric power consumers was the continuity of supply or their reliability. Nevertheless, nowadays network users demand a product quality that must meet all standard requirements and customer expectations in a scenario where renewable energy is introduced. In the previous chapter, it was shown that this is a framework that is currently being defined.

Power Quality Measurement and Analysis Using Higher-Order Statistics: Understanding HOS Contribution on the Smart(er) Grid, First Edition. Olivia Florencias-Oliveros, Juan-José González-de-la-Rosa, José-María Sierra-Fernández, Manuel-Jesús Espinosa-Gavira, Agustín Agüera-Pérez, and José-Carlos Palomares-Salas.
© 2023 John Wiley & Sons Ltd. Published 2023 by John Wiley & Sons Ltd.

The quality of the product – from now, the quality of the waveform – involves companies in the electricity sector at different levels: generation, transmission and distribution, as well as those related to the users of the network (i.e. end-customers and prosumers) and equipment manufacturers. There is a current trend for industrialised countries to have established standards that limit the disturbance behaviours in electrical systems, paying special attention to some sectors and components involved, such as the manufacturing industry, with the introduction of most controlled processes through the Internet of Things (IoT). Responsibility and penalisation have been established in the standards between stakeholders.

Poor waveform quality is a problem because it involves energy losses, a non-efficient performance in most additional loads, an overload in transport networks and economical losses due to non-efficient production or even a lack of it. Certainly, in order to achieve the essential network characteristics, it would be necessary to preserve the sinusoidal voltage waveform from its production, in both power flows, from the utility and from the end-user installations, and throughout all the system. Nevertheless, as mentioned in the first chapter, nowadays most PQ waveform problems are to be found within medium voltage (MV) to low voltage (LV) installations. The measures that are included in this work contemplate monitoring network behaviour at LV [69].

2.2 Background on Power Quality

According to IEC 61000-4-30 [1], PQ is defined by the characteristics of the electricity at a given point on an electrical system, evaluated against a set of reference technical parameters. These parameters might, in some cases, be related to the compatibility between electricity supplied on a network and the loads connected to that network. They are: magnitude, distortion, symmetry, phase, frequency, the displacement factor and the like.

The quality of the voltage waveform can be studied from different perspectives, depending on the energy received from the utility and the quality of the voltage within customer facilities (see Appendix A, Voltage Waveform) relative to the theoretical power system waveform. Indeed, the network impedance usually has a significant influence on such behaviour.

Nevertheless, there are various types of loads connected to the network, among which are the linear and non-linear loads. Indeed, linear loads also cause sinusoidal currents with the same frequency as the voltage, although there could be a gap between the two waveforms (i.e. voltage and current). However, with the development of power electronics, plenty of products have been introduced in the network because of their high-reliability and low-cost features. Despite of this, new power electronic solutions can represent distorting loads, also known as non-linear from a point of view of the electrical system. The introduction of non-linear loads in the grid will continue to increase. It is known that non-linear loads introduce distorted currents in the network, which can reach a considerable amplitude, depending on the network impedance. The combined effect of the multiple non-linear loads has an impact on the voltage waveform [70]. This affects the performance of the facilities. The presence of these new loads is directly related to the quality of the waveform in the power system.

Efforts in PQ and energy efficiency fields are directly focused on characterising the behaviour of these loads and electronic devices connected to the system. For example, on the replacement of conventional lighting by light-emitting diode (LED) lighting technologies, the waveform quality is monitored to characterise harmonic emission of new solutions [71]. Nevertheless, harmonics are not the only source of defects as many more kinds of events are caused when devices are connected or disconnected.

Despite this, the current quality of voltage waveform is not constantly measured in the distribution networks, which this is one of the main challenges that researchers and instrumentation enterprises must face in the smart grid (SG). Indeed, this is one of the motivations of the present work characterising the distribution of the waveform from the time domain with a statistical perspective.

In the present context, it is important to characterise both the voltage and the current wave shape. Despite this, current characterisation from the time domain is more difficult to define than voltage because it changes constantly. Moreover, PQ standards are mostly devoted to voltage characterisation. From a statistical point of view, it is easier to establish a methodology for studying the voltage waveform.

Energy quality indices, as happens with all engineering indicators, quantify the information and compress it into a single parameter. However, they can hide something behind what they actually show. Most of the indices summarise the waveform distortion of a sinusoidal voltage. Such distortion may come from communication causes, the telephone and other coupled phenomena [71]. For all these reasons, in the future grid, PQ analysis and evaluation will be needed before taking any corrective action.

2.3 Traditional Theories of Electrical Time Domain

There are well-established traditional techniques that are applied in electrical instrumentation as waveform characterisation. Common time-domain indices are root-mean-square (RMS) and, for a nominal voltage reference, the peak value (PV). Some instruments such as multimetres and oscillo-scopes have introduced other indices, such as the form factor (FF), crest factor (CF), peak method or the average responding value on a cycle.

2.3.1 RMS Value

The RMS value is an electrical magnitude that correlates a periodic wave-form and helps to detect rapid voltage changes in the time domain. However, even when the RMS does not measure the voltage waveform shape it is deduced/inferred that a sinusoidal voltage exhibits an RMS value of 0.707.

Equation (2.1) shows the RMS calculation:

$$f_{RMS} = \lim_{T \to \infty} \sqrt{\frac{1}{T} \int_0^T \left[f(t) \right]^2 dt} \tag{2.1}$$

where T is the period of the signal.

According to reference [1], for a class A instrument, the detection of rapid voltage changes (RVCs) should be initiated as a recording of an initial set of $100/120\ U_{rms}$ (1/2) for 50/60 Hz systems, respectively, and computing the arithmetic mean of those values.

In addition, in a performance analysis there is a requirement to measure the RMS voltage in a range between 1 and 10 minutes.

2.3.2 Peak Value

For a sinusoidal waveform, the peak value is the maximum value that must coincide with the maximum voltage. If the signal is sinusoidal and the RMS is 0.707, then the peak value is 1. The handicap is that the information provided by the peak value is confused if the voltage signal has a more complex form.

2.3.3 Form Factor Value

The FF value is the ratio between the RMS of a waveform and the absolute average (i.e. the rectified average) of the waveform and it is a convenient way to refer to distortion of the waveform as well as its heating effect. The theoretical value for a rectified sinusoidal waveform is 1.1.

Mathematically, it can be expressed as

$$\text{Form Factor} = \frac{V_{RMS}}{V_{av}} \tag{2.2}$$

The indicator helps in the characterisation of a rectified waveform, although from a mathematical point of view it is impossible to summarise the FF because the average of a non-rectified sinusoid is null and the division by zero does not have a defined value.

2.3.4 Crest Peak Value

The crest peak value is defined as the ratio of the maximum value to the RMS value of an alternating quantity. The theoretical value for a sinusoidal waveform is 1.41.

It can be expressed as

$$\text{Crest factor} = \frac{V_{max}}{V_{RMS}} \tag{2.3}$$

The CF is a measure of the degree of sinusoidal conditions in lab measurements aimed to test different loads. In that case, the devices are feeds with perfect waveforms that must exhibit a CF of 1.41.

Nevertheless, even when some of these indices can exhibit a classification related to voltage waveform, they come from the instrumentation world and are not aligned to PQ measurements in the standards IEC 61000-4-30, except the RMS. However, in order to relate the HOS classification with an electrical measurement, they will be mentioned as some explanations are given in this chapter.

2.4 HOS Contribution in the PQ Field

Different HOS estimates have been proposed throughout the last few decades to infer new statistical characteristics associated with data from a non-Gaussian time-series in a predominantly Gaussian background, which can theoretically be considered as a result of the summation of different noise processes.

Within the PQ disturbance detection context, as previously discussed, the targeted electrical disturbance is always considered to be non-Gaussian, while the floor is assumed to be a stationary Gaussian signal. Thus, using HOS would help to locate the perturbation qualitatively, with a rough approximation, but allowing detection not only in specific measurements but also in a continuous measurement campaign.

Continuous PQ phenomena, such as frequency fluctuations or discrete items, such as sags, swells, and transient events, are difficult to model under a real framework because they appear as multiple events most of the time. Indeed, the modern electrical network demands a new approach in order to detect various phenomena [37].

2.4.1 HOS Indices Definitions

A summary of the equations of the cumulants of the centred moments (from the first to fourth one), which are the indices based on HOS used in this work, can be found in Appendix B, Time-Domain Cumulants. An interpretation about the potential use of each cumulant in PQ applications is introduced below.

Variance (v) measures the degree of dispersion around the mean; positive and negative deviations contribute equally, resulting in a positive variance value. In power supply signals, variance helps to detect changes in the signal amplitude, which can be indicative of sag and swell events. In addition, variance helps to characterise the power supply under continuous normal operating conditions.

Skewness (s) is a measure of the degree of symmetry of data distribution. Skewness can be positive or negative depending on the sizes of the right and left tails of the distribution. If the left (right) is more evident than the right (left), the skewness is negative (positive). In PQ events, non-symmetry is generally a sensitive parameter that indicates half the cycle in which the event has taken place. This characterisation can be useful in order to detect transients, the initial and final half-cycles in sags and swells.

Kurtosis (k) characterises the tails of statistical distribution. In the bimodal distribution of a voltage sinusoidal cycle, the tails of the distribution are the maximum and minimum values. Indeed, the kurtosis measures the outliers – in the form of heavy tails – mainly in the region of the maximum and the minimum values of the wave shape. In the context of voltage waveforms, the flatter the top and down regions, the lower the kurtosis will be.

The theoretical variance, skewness and kurtosis of an ideal voltage supply of 50 Hz (Figure 2.1) are variance ($v = 0.5$), skewness ($s = 0$) and kurtosis ($k = 1.5$). These respective values help to characterise a bimodal distribution function through the HOS domain.

2.4.2 HOS Performance in Signal Processing

The robustness of HOS has been tested before in order to make a voltage characterisation related to some signal processing aspects. HOS performance detection is summarised in the Tables 2.1–2.4. Measurements were

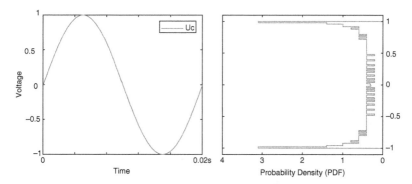

Figure 2.1 Theoretical 50 Hz power supply and the PDF that characterises a bimodal distribution from a standard sinusoidal nominal supply voltage (U_c). *Source:* Olivia Florencias-Oliveros taken from reference [69].

Table 2.1 Detector performance due to the signal-to-noise ratio.

SNR (dB)	20	25	30	40	50
True (%)	83	62	100	100	100
Error (%)	17	38	0	0	0
Undefined	0	0	0	0	0

Source: Authors.

Table 2.2 Detector performance due to the frequency sampling.

F_s (kHz)	4	5	10	15	20	25
Parts/cycle	80	100	200	300	400	500
True (%)	100	100	100	100	100	100
Error (%)	0	0	0	0	0	0
Undefined	0	0	0	0	0	0

Source: Authors.

Table 2.3 Detector performance due to the number of cycles computed through HOS.

F_s (kHz)	1	2	3	4	5	10
True (%)	100	100	100	100	100	100
Error (%)	0	0	0	0	0	0
Undefined (%)	0	0	0	0	0	0

Source: Authors.

Table 2.4 Detector performance related to the fundamental frequency fluctuations established by the standard [13].

	Range for isolated networks		Range for synchronous networks	
F_s (Hz)	48	52	49.9	50.1
True (%)	100	100	100	100
Error (%)	0	0	0	0
Undefined (%)	0	0	0	0

Source: Authors.

analysed over a set of 1000 synthetic signals according to the IEEE 1159 Working Group.

HOS have shown their robustness under Gaussian noise, as concluded in some publications in Section 1.4.9 and summarised in Table 2.1. Up to 30 dB seems to be a convenient range for the measurement using the HOS algorithms. However, the power system has a signal-to-noise ratio (SNR range) between 50 and 70 dB, from which the robustness of the HOS seems to be more than convenient. All the measurements in Table 2.1 were made at a frequency sampling of 20 kHz.

Table 2.2 shows that, according the Nyquist theorem, the minimum frequency sample used for processing the 50 Hz voltage waveform with the HOS indices is 5 kHz. Nevertheless, in all the measurements made through this document, synthetics and real-time measurements have been processed at a frequency sample of 20 kHz, as PQ standards recommend [72].

The HOS robustness allows the measurement strategy to be extended according to measurement objectives scaling the sliding window from 1 cycle to 10 cycles (10 T), as the IEC 61000-4-30 suggests. Notice that in Table 2.3 the number of computed cycles does not impact HOS measurements for a steady-state behaviour. Sliding window strategies used in this document are adapted to the application needs:

– Short sliding window for HOS feature extraction: 1 cycle (1 T).
– Long sliding window for HOS feature extraction: 10 cycles (10 T).

HOS detect fundamental frequency variations, which helps to characterise frequency deviations in a 50 Hz power system. The characterisation of the frequency deviation is discussed here. As observed in Table 2.4, the accuracy of the measurements under fundamental frequency changes is shown.

2.4.3 HOS vs. Electrical Time-Domain Indices

As concluded in Section 2.3, the time-domain indicators from which a relationship can be established with HOS are RMS, CF and peak factor. In this sense, amplitude changes under different sinusoidal voltages are represented in Table 2.5. The first steady-state is a nominal supply voltage (U_c) normalised, while the second and third signal states present supply voltage variations ($U_{din} \pm 0.1$). The variance infers the behaviour of some other electrical magnitude changes in the voltage waveform; for example, RMS behaviour. Nevertheless, the CF does not detect a relevant change in the waveform. That is why the variance seems more convenient to use in feature extraction characteristics.

Figure 2.2a shows the same amplitude changes under different sinusoidal voltages but with an increasingly cycle-to-cycle behaviour. The probability density function (PDF) of the three waveforms is equal, a bimodal distribution (Figure 2.2b). Notice that, even when the amplitude changes $\pm 10\%$ Uc, the distributions in Figure 2.2b keep the same characteristics: sinusoidal and symmetrical.

In Figure 2.2c and d, amplitude changes are computed in variance vs. V_{rms} and the CF, respectively. The theoretical values are indicated by dotted lines. Notice that RMS exhibits a linear relationship with the variance that computes the changes in the mean as consecutive changes when increasing and decreasing the voltage amplitude. Changes in the V_{max} are computed by the variance and the RMS. The CF mathematically depends on the V_{peak}, V_{max} and RMS. Less amplitude in the voltage is computed using the lowest variance, RMS, CF and vice versa. The analysis in Figure 2.2 helps to validate the convenient use of variance in time-domain studies.

The last analysis in Figure 2.2 is useful in order to show the high sensitivity of the variance index for waveform characterisation.

Table 2.5 HOS vs. electrical time domain indices.

Voltage	HOS		
	Variance	RMS	Crest factor
Normalised U_c	0.5	0.707	1.41
U_{din} 0.9	0.4060	0.670	1.415
U_{din} 1.1	0.6065	0.745	1.415

Source: Olivia Florencias-Oliveros taken from reference [69].

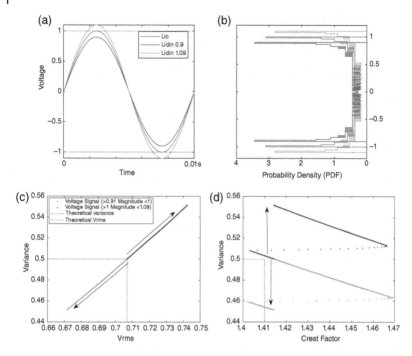

Figure 2.2 Variance vs. other electrical parameters under amplitude changes of the normalised U_c, U_{din} 0.9 and U_{din} 1.1 power supply. (a) Cycle-to-cycle amplitude changes in the power supply. (b) PDF of the power supply under cycle-to-cycle amplitude changes. (c) V_{rms} vs. variance under cycle-to-cycle amplitude changes. (d) Crest factor vs. variance under cycle-to-cycle amplitude changes. *Source:* Olivia Florencias-Oliveros taken from reference [69].

2.5 Regulations

The UNE-EN50160 standard defines, describes and specifies the voltage supply characteristics that are the main objectives concerning the PQ measurements, some of them being amplitude changes, frequency deviations, waveform shape and symmetry at different voltage levels. Indeed, all the characteristics are subject to variations during normal operation in a power system due to load modifications, disturbances emitted by some equipment and the occurrence of faults, mainly from an external origin.

Power supply characteristics vary randomly and depend on the monitored power supply point and also randomly in space with respect to a given instant. In fact, the values established by standards can vary due to the random nature of the power system. Nevertheless, the standard UNE-EN 50160 does not regulate the fulfilment responsibilities by either party (i.e. supplier and end-customer). The standard is only applied under normal operating conditions.

Furthermore, IEC 61000-4-30 describes the established measurement methods in voltage supply studies for the following parameters: frequency, amplitude of the voltage supply, flicker, sags, temporary overloads, voltage interruptions, transients, voltage imbalance, voltage harmonics and inter-harmonics, network transmission signals and RVCs.

The proposed method does not seek to align the entire study to the current normative, for this statement has already been discussed in Chapter 1. This document aims to explore the ability of HOS to respond to actual and future PQ problems. In future PQ studies, the focus should be changed to incorporate a different approach to measure the network with a statistical perspective.

2.6 The Sliding Window Method for HOS Feature Extraction

The proposed method uses the sliding window, where instantaneous values are compared with the corresponding values on previous cycles in the time domain. The study is mainly focused on quantifying the main distortions or deviations of the power supply through HOS results that are summarised in both standards, UNE-EN50160 and IEC 61000-4-30, in a time-domain analysis under dynamic conditions, such as:

– Amplitude: to detect changes in the magnitude of the waveform in the time-domain.
– Frequency: to detect changes in the fundamental frequency with a difference of 50 Hz.
– Waveform shape: to measure symmetry and waveform deviations with a 50 Hz power supply.

The minimum requirements of the system are:

– Frequency sampling (F_s) = 5 kHz.
– Signal-to-noise ratio (SNR): minimum of 30 dB.
– Zero-pass detection to initial signal capture.

All deviations are related to a PQ disturbance. Deviations are measured in different intervals according to PQ studies. Both PQ characterisation phenomena, monitoring in continuous and detecting events, need to be known in advance, as well as the measurement uncertainty and the HOS method capabilities, in terms of detection and classification. Indeed, in this chapter HOS performance is introduced with the objective of summarising all their capabilities as well as the impact that it would have on further applications based on these statistics.

In addition, a PQ index based on HOS is introduced as an indicator that computes deviations that would help to obtain a holistic analysis of the waveform state. The index computes the waveform under three criteria: non-symmetrical, non-sinusoidal and waveform deviations. The objective is to explore whether these three main states can be characterised in more advanced studies.

2.6.1 Amplitude Changes

In PQ standards, amplitude changes are focused on detecting conformity areas under different reference amplitude values, which are related to deviations of the theoretical power supply regarding nominal voltage. Quantifying voltage amplitude changes is important in order to test the network state and different monitoring objectives, as well as to detect sags, swells, rapid voltage changes, transients and other events, even those that have not been characterised.

Table 2.6 shows the theoretical values as results of detecting amplitude changes in U_{din} 0.9 and U_{din} 1.09. Indeed, normal operating conditions given in voltages remain within this range.

According to measurements in Table 2.6, the variance detects the amplitude changes while skewness and kurtosis remains in their theoretical values of $s = 0$ and $k = 1.5$. That is indicative of a sinusoidal behaviour according to the waveform analysed in Figure 2.2a. Nevertheless, skewness and kurtosis change under other waveform shape deviations. This is something that has been discussed in the sag and swell detection paper [73].

In Figure 2.3, the PDF of different statistics shows that their ranges are computing a sinusoidal waveform that changes their amplitude cycle-to-cycle between a voltage range (U_{din}) from 0.9 to 1.09.

Table 2.6 HOS approach for defining PQ indices based on the waveforms in Figure 2.2a.

Voltage	HOS		
	v	s	k
Normalised U_c	**0.5**	**−4.1518e-16**	**1.5**
U_{din} 0.9	0.4560	1.2361e-16	1.5
U_{din} 1.1	0.6065	2.0813e-16	1.5

Source: Olivia Florencias-Oliveros taken from reference [69].

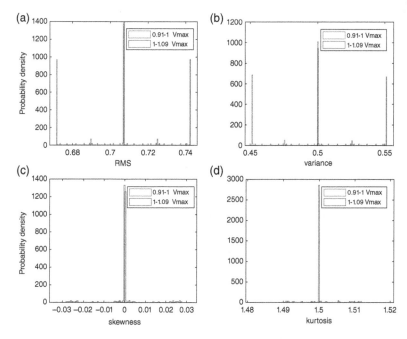

Figure 2.3 PDF of the different indices and their individual ranges under different voltage amplitude (U_{din} from 0.9 to 1.09). (a) PDF of the RMS under cycle-to-cycle voltage amplitude variations. (b) PDF of the variance under cycle-to-cycle voltage amplitude variations. (c) PDF of skewness under cycle-to-cycle voltage amplitude variations. (d) PDF of kurtosis under cycle-to-cycle voltage amplitude variations. *Source:* Olivia Florencias-Oliveros taken from reference [69].

2.6.2 Phase Angle Jumps

It should be remarked that the cycle-to-cycle phase-angle jumps can be detected using HOS. This can be observed in Figure 2.4, where different two-dimensional (2D) graphs (a to d) represent the two states based on the amplitude changes and the phase-angle jump deviations as a result of the transition states in the synthetic waveform. The arrows indicate the direction in which the amplitude increases and decreases in the statistical 2D planes (Figure 2.5).

Figure 2.4 proves that non-stationary and non-symmetrical waveforms can be detected from the sliding windows analysis and the 2D HOS planes. This is demonstrated in the evolution of the simulation through phase-to-cycle phase shifts as a result of non-symmetry behaviour. In addition, different steady states are detected. Moreover, the simulation helps to establish a conformity region within the voltage supply variations that, following standard EN 50160:2010/A1:2015, must be within ±10% of the nominal voltage in all power terminals of the LV network 100% of the time. The ranges are summarised in Table 2.7.

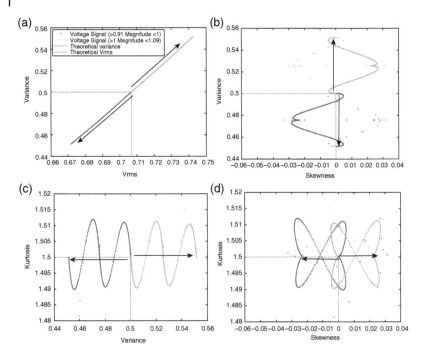

Figure 2.4 Evolution of the individual statistic ranges under cycle-to-cycle voltage amplitude variations in the HOS planes. (a) Variance vs. RMS under cycle-to-cycle voltage amplitude variations. (b) Skewness vs. variance under cycle-to-cycle voltage amplitude variations. (c) Variance vs. kurtosis under cycle-to-cycle voltage amplitude variations. (d) Skewness vs. kurtosis under cycle-to-cycle voltage amplitude variations. *Source:* Olivia Florencias-Oliveros taken from reference [69].

2.6.3 Fundamental Frequency

Fundamental frequency changes are established by the standard EN 50160:2010/A1:2015. In systems with a synchronous connection to an interconnected system, they should normally be within 50 ± 1 Hz for 100% of the time. Meanwhile, for systems that do not have a synchronous connection to an interconnected system, the frequency must be within 50 ± 2 Hz for 100% of the time.

In Figure 2.5, a simulation has introduced fundamental frequency changes in a signal that increases from 50 to 50.09 Hz. The objective is to show that HOS are capable of computing the instantaneous deviations in frequency as observed in Figure 2.5a. In addition, in the HOS planes, the frequency transition states are detected as trajectories such as those in Figure 2.5b to d.

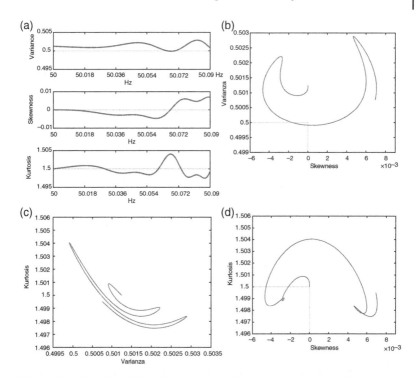

Figure 2.5 Evolution of the HOS trajectories for a voltage waveform that increases the fundamental frequency cycle-to-cycle from 50 Hz to 50.09 Hz. (a) Cycle-to-cycle evolution of HOS under frequency changes ±1 Hz. (b) Cycle-to-cycle evolution of skewness vs. variance under frequency changes ±1 Hz. (c) Cycle-to-cycle evolution of variance vs. kurtosis under frequency changes ±1 Hz. (d) Cycle-to-cycle evolution of skewness vs. kurtosis under frequency changes ±1 Hz. *Source:* Olivia Florencias-Oliveros taken from reference [69].

Table 2.7 HOS ranges that summarise voltage supply variations of ±10%.

	HOS		
Voltage Supply variations	v	s	k
Normalised U_c	**0.5**	**−4.1518e-16**	**1.5**
U_{din} 0.9–1.1	0.45–0.55	(−0.03)-0.03	1.49–1.512

Source: Olivia Florencias-Oliveros taken from reference [69].

Furthermore, based on the results obtained in Section 2.6.1, the fundamental frequency remains within the conformity region related to power supply amplitude variations. In Figure 2.6, the same signal is processed, although

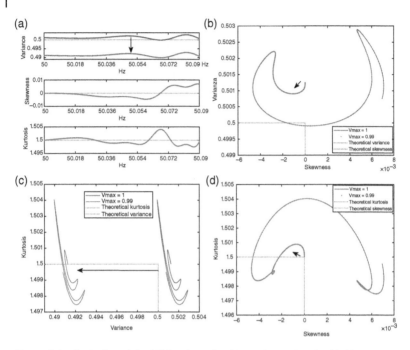

Figure 2.6 Evolution of the HOS trajectories for a voltage signal with U_{din} between 0.99 and 1, which increases the fundamental frequency cycle-to-cycle from 50 Hz to 50.09 Hz. (a) Cycle-to-cycle evolution of HOS under frequency changes ±1 Hz, U_{din} 0.99. (b) Cycle-to-cycle evolution of skewness vs. variance under frequency changes ±1 Hz, U_{din} 0.99. (c) Cycle-to-cycle evolution of variance vs. kurtosis under frequency changes ±1 Hz, $\underline{U}_{din}$ (0.99–1). (d) Cycle-to-cycle evolution of skewness vs. kurtosis under frequency changes ±1 Hz, U_{din} (0.99–1). *Source:* Olivia Florencias-Oliveros taken from reference [69].

with less amplitude (U_{din} 0.99); the variance detects the amplitude deviation while skewness and kurtosis remain in their theoretical values.

Irregular frequency changes pose a problem for some industries and impact the performance of different devices. An increasing amount of equipment demands the highest frequency protection. System operators are devoted to controlling large fundamental frequency deviations that cause massive equipment tripping. Small fundamental frequency variations might impact control algorithms of certain control power electronic converters. This issue is an EMC question, even though it is important that the method can compute the frequency deviations instantaneously.

The fundamental frequency uncertainty in HOS planes has been computed. More information can be found in Chapter 3, Section 3.5, about individual statistics ranges for isolated networks and synchronous networks.

2.6.4 Waveform Shape Deviation

The signals analysed in this chapter have been modelled as sinusoidal. However, measurements taken in the field confirm that the power supply exhibits the probability of not being completely ideally sinusoidal. Multiple aspects can influence their characteristics, such as network type, monitoring point, space and time. The small differences in amplitude, symmetry and sinusoidal aspects could be computed through an HOS feature extraction analysis.

For a better understanding of HOS analysis in the field, two cycles from different networks have been characterised (see Figure 2.7) compared to the theoretical power supply. The theoretical voltage waveform is characterised through the theoretical HOS, variance ($v = 0.5$), skewness ($s = 0$) and kurtosis ($k = 1.5$). The voltage supply and PDF of two waveforms shown in Figure 2.7, where Case 1 is from the UCA (University of Cádiz) and Case 2 is a household signal. Both waveforms are computed through the individual HOS, where in Case 1 $v = 0.4826$, $s = -0.0067$, $k = 1.5092$ and in Case 2 $v = 0.4720$, s $= -0.0039$, $k = 1.4726$.

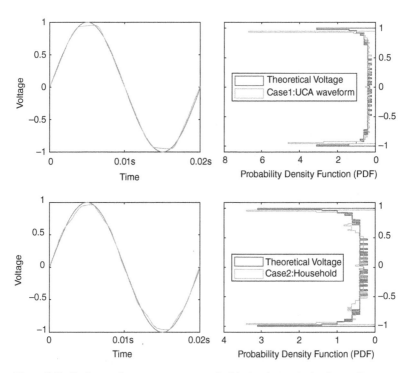

Figure 2.7 Both waveform cases, compared with the theoretical voltage. *Source:* Olivia Florencias-Oliveros taken from reference [69].

Both signals remain within the conformity region of voltage supply variations. Two different degrees of deviation can be detected. In the first signal, supplied by UCA, the variance detects the amplitude change, because the signal exhibits a flat-top behaviour, while skewness and kurtosis remain around their ideal values. Typical voltage distortions such as flat-top are the result of non-linear loads from single-phase rectifiers and different power electronic emissions in LV networks.

Focusing on the second case, the household, the deviations in the top and down intervals denote certain levels of harmonic distortion that is detected by the kurtosis. A more pointed top waveform is computed by a lower kurtosis. Also, as both cycles are symmetric, the deviation in the skewness is negligible.

According to the time-domain information from the PDF and HOS results, the most deteriorated signal comes from the household network. The voltage distortion is measured following the criterion of the waveform deviation from the ideal voltage or, in other words, a contribution of the original frequency regarding its harmonics. Thus, quantifying this small deviation is the result of relating this value with a deviation in the harmonics that contains this signal in the frequency domain (see Figure 2.7). Therefore, Figure 2.8 quantifies HOS deviations of both steady-state voltage waveforms from the theoretical state.

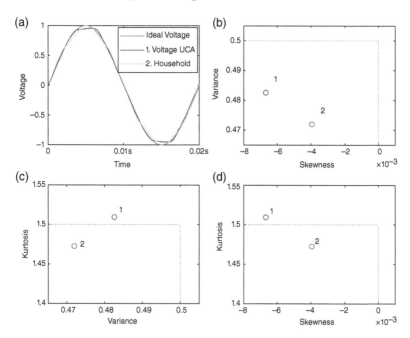

Figure 2.8 HOS deviations of both steady-state voltage waveforms in Figure 2.7 from the theoretical state of an ideal voltage. (a) Steady-state voltage waveforms. (b) Skewness vs. variance steady-state for both signals. (c) Variance vs. kurtosis steady-state for both signals. (d) Skewness vs. kurtosis steady-state for both signals. *Source:* Olivia Florencias-Oliveros taken from reference [69].

Notice that both waveforms in Figure 2.7 remain within regular operating conditions. In the next chapters, the contribution of HOS feature extraction under different approaches, as well as PQ monitoring objectives, will be outlined.

2.7 PQ Index Based on HOS

In order to characterise the undeniable complexity of the current electricity grid, it can be convenient to use a measurement index based on HOS that would detect new types of electrical disturbances and multiple disturbances (i.e. superposition of two or more events).

The PQ index based on HOS establishes a vector that computes the three main characteristics of an ideal 50 Hz power supply that is symmetrical and sinusoidal in terms of shape. The main objective of this strategy is to monitor the quality of the waveform, based on different steady-state compliance regions. The proposed PQ indicator for permanent monitoring focuses on dealing with the large amount of data related to online analysis.

Learning the stationary theoretical state and computing the deviations from that state helps to characterise the waveform pattern under the contractual operating conditions of the analysed network. Additionally, the information relative to the quality of the waveform helps to simplify future monitoring campaigns using the advance monitoring infrastructure with less computation effort and resources from the very beginning.

The deviations in the ideal power supply are always related to waveform changes; each kind of event is associated with a degree of shape deviation [58]. However, the majority of the parameters introduced in the regulations only take into account the energy of the system, such as the traditional indices V_{rms} or total harmonic distortion (THD). In most cases, the time-domain information is not recoverable and remains difficult to interpret [62]. In fact, the PQ index based on HOS provides additional information that may be useful when compared with traditional indices and thus it is possible to calculate the result of the energy changes that are reflected in the deviations of the voltage waveform. In this sense, a methodology adapted to a PQ index based on HOS must first be developed in order to use the indicator in a continuous monitoring campaign.

The contribution of the individual statistics to the PQ index is related to the ability of HOS to add information that helps to characterise the distribution of the voltage waveform within a single indicator. As mentioned before, the theoretical values for the variance, skewness and kurtosis corresponding to an ideal voltage supply of 50 Hz are 0.5, 0.0 and 1.5, respectively. This triplet is assumed to be the steady-state values from which to measure deviations on the HOS planes and compute the deviation index. While the theoretical value of the index is zero, in real practice it has to be calibrated depending on the point under test, as well as the specific operating conditions that converge at this point under test. In order to define the PQ index, the following magnitudes should be taken into account:

Δt = measurement interval
s_{ij} = jth statistic associated with the ith period
$\hat{s}_j$ = nominal jth statistic
M = number of periods contained in Δt
N = number of statistics for characterisation

Given a measurement point in the SG, realisation of the PQ deviation index is a function of the specific deviations of each individual statistic with respect to their nominal values, and is given by the generic expression

$$PQ_{\Delta t} = f\left(\left|s_{11} - \hat{s}_1\right|, \ldots, \left|s_{ij} - \hat{s}_j\right|, \ldots, \left|s_{MN} - \hat{s}_N\right|\right) \tag{2.4}$$

A particular case of Eq. (2.4) consists of using the summation of each individual deviation and is presented as

$$PQ_{\Delta t} = \frac{\sum_{i=1}^{M}\sum_{j=1}^{N}\left|s_{ij} - \hat{s}_j\right|}{M} \tag{2.5}$$

The deviations of each statistic from their ideal values assess and quantify the waveform. The present research uses three deviation terms in Eq. (2.5) and the final expression for the index is described as

$$PQ_{\Delta t} = \frac{\sum_{i=1}^{M}\left|var_i - v\hat{a}r\right| + \left|sk_i - \widehat{sk}\right| + \left|kur_i - k\hat{u}r\right|}{M} \tag{2.6}$$

where *var*, *sk* and *kur* represent variance, skewness, and kurtosis, respectively.

The proposed PQ index is used to compute the different strategies in the previous analysis. In Figure 2.9, the cycle-to-cycle PQ index summarises amplitude changes relative to the signal in Section 2.4.1. The maximum values are around 0.05 and the phase-angle jumps achieve a PQ of 0.07.

As the indicator operates in absolute values, it computes as the same deviation, amplitude increase or a decrease in the equal magnitude proportion. In addition, fundamental frequency changes from Section 2.6.3 are computed by the index in Figure 2.10. Moreover, amplitude changes have an impact on the index with a relationship where $\pm 1\,$Hz of deviation achieves a

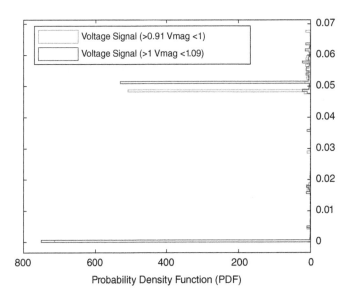

Probability Density Function (PDF)

Figure 2.9 PQ index and the PDF evolution that compute the cycle-to-cycle (0.02 s) amplitude changes from Figures 2.2 and 2.3. *Source:* Olivia Florencias-Oliveros taken from reference [69].

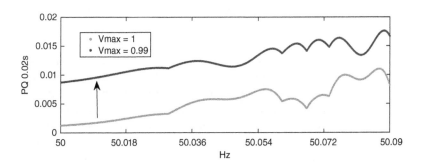

Figure 2.10 PQ index that computes simulated frequency changes and amplitude deviations following the measurements in Sections 2.6.1 and 2.6.3. *Source:* Olivia Florencias-Oliveros taken from reference [69].

value of 0.01. It should be highlighted that the increase in the index is given by an increase in the amplitude, which is mainly derived by variance.

Such an index helps a degree of deviations in field measurements to be detected, characterising different steady states. In Figure 2.10, both cases belong to increasing frequency changes and amplitude deviations conducted by measurements taken in Sections 2.6.1 and 2.6.3, where in Case 1 the initial PQ = 0.0333 and in Case 2 PQ = 0.0593. Kurtosis helps quantify

Table 2.8 Maximum values achieved by the index under different steady-state conditions.

Steady-state conditions	PQ index based on HOS
Amplitude changes (U_{din} 0.9 − U_{din} 1.09)	0.068
Households-pointed	0.037
UCA-flat-top	0.058
Amplitude and frequency changes 0.02 s	0.018
Frequency changes 0.02 s	0.01

Source: Olivia Florencias-Oliveros taken from reference [69].

a non-sinusoidal state, as in the second case, thus showing that it is the most deteriorated signal.

Finally, measurements help to identify different PQ thresholds in Table 2.8 from previous measurements. The indicator is sensitive to cycle-to-cycle amplitude deviations, fundamental frequency and a steady state based on the quality of the waveform shape.

2.8 Representations Used by the Time Domain

In order to characterise the undeniable complexity of the current electricity grid, it can be convenient to use a measurement index based on HOS, which would detect new types of electrical disturbances and multiple disturbances (i.e. superposition of two or more events). Throughout this dissertation, different graphs relating to HOS measurements are represented in bi-dimensional planes. Necessary knowledge about the representations used by the proposed methods is introduced and explained and different strategies are explored in the next chapters using these graphs.

HOS representation as 2D planes helps to compute the waveform cycle-to-cycle changes as a map where the origin of coordinates is based on a theoretical HOS of a 50 Hz power supply. The 2D planes also help to represent computed HOS behaviour based on their trajectory (see Figure 2.11).

As the time-analysing window sweeps the signal under test, 2D planes are generated, and the following points are considered. Each point in a 2D-HOS plane (a pair of coordinates) corresponds to one cycle in the time domain; this is precisely the temporal resolution of the measurement method and the N:1 time compression. The time compression relationship will be adapted according to measurement objectives.

The steady-state triplet is given by variance = 0.5, skewness = 0.0 and kurtosis = 1.5, which describes the non-disturbed ideal voltage waveform

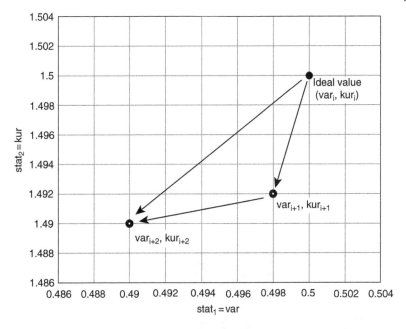

Figure 2.11 Bi-dimensional HOS plane, for example variance (var) vs. kurtosis (kurt), while certain disturbance is analysed cycle-to-cycle. Theoretical two-dimensional HOS traces taken from reference [74]. *Source:* Authors.

located in the origin of each coordinate in the three 2D-HOS planes. The deviations of each statistic from their ideal values assess and quantify the waveform instantaneously. The strategy of measuring trajectories is convenient for characterizing PQ events. Nevertheless, when a long-time series is analysed, other representation techniques are more convenient in order to appreciate variations on HOS planes when more information is computed. It is essential that the chosen method allows us to detect and interpret variations without affecting other aspects related to measurements.

In this sense, the waveform shape deviation fingerprint (2D planes) helps us to scale the persistence of a pairwise vector (see Figures 2.12–2.14). The matrix is the temporary evolution of a deviation (x, y) vector that can compute cycle-to-cycle variance, skewness and kurtosis behaviour.

The matrix can be adapted for detecting both normal operation conditions and events. In the colour scale, z axes show a percentage and indicate the different persistent regions according to the measurement temporary window, which is aligned to temporary monitoring objectives (in scalable time it is possible to analyse from 0.02 s up to a week or more). Using these 2D planes (named as waveform-shaped deviation fingerprints), variations between time intervals are highlighted, thus showing the evolution of the signal pattern.

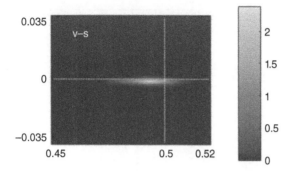

Figure 2.12 Fingerprint of the waveform time-series deviation in the HOS planes, variance vs. skewness (v–s). Each time interval is characterised by its associated PQ, index = 0.0116. *Source:* Authors.

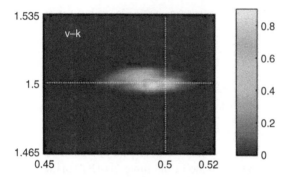

Figure 2.13 Fingerprint of the waveform time-series deviation in the HOS planes, variance vs. kurtosis (v–k). Each time interval is characterised by its associated PQ, index = 0.0116. *Source:* Authors.

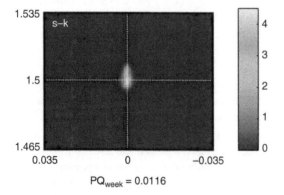

Figure 2.14 Fingerprint of the waveform time-series deviation in the HOS planes, skewness vs. kurtosis (s–k). Each time interval is characterised by its associated PQ, index = 0.0116. *Source:* Authors.

It should be noted that colours do not indicate any excess threshold, but the area of greater persistence during the sampling interval. Each colour represents the differences between a certain value and the maximum of all measurements in each scenario. The colour scale varies from 0%, which is the furthest, to the maximum value 100%, which represents a signal that has been in this steady-state region for the most time within the monitoring interval. As an example, the three matrices are shown for the same measurement interval. Notice that the different regions are highlighted, where

M = symmetry vs. sinusoidal
N = symmetry vs. amplitude
O = amplitude vs. sinusoidal

The HOS plane representations are helpful in explaining a scalable proposal. In Figure 2.15, the former concepts associated with the method

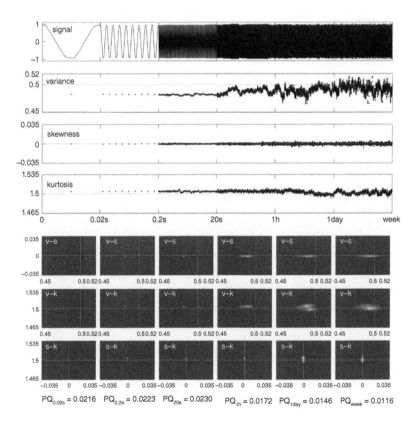

Figure 2.15 One-week monitoring of one week using the method based on HOS with a minimum resolution of one cycle. *Source:* Authors taken from reference [75].

are explained. The temporal analysis has been segmented into six main intervals or observation periods (from left to right), each of which includes the previous one, but overlapped. In the very first period of the series, it can be appreciated that the sinusoid waveform is not ideal at all. In the following 10 cycles, the defects begin to be hidden, which are then followed by the sequences: 20 seconds, 1 hour, 1 day, and 1 week segments. The complete one-week record is located at the top of Figure 2.15. From top to bottom are shown the complete power supply time series, variance evolution, skewness evolution, kurtosis evolution, the variance-to-skewness graph family, variance vs. kurtosis 2D graphs and a skewness vs. kurtosis set of graphs. Each time interval is characterised by its associated PQ index.

3

Event Detection Strategies Based on HOS Feature Extraction

3.1 Introduction

This chapter summarises different experiments focused on aligning the higher-order statistics (HOS) measurement strategies for the most suitable disturbance detection in future networks. The measurements were taken in the Electronic Laboratory of the University of Cádiz. Events detection, such as sags, fundamental frequency changes and harmonics, is introduced here. Individual strategies are shown in order to develop off-line measurement and monitoring solutions based on HOS. All this information is relevant in order to introduce event detection in the field in comprehensive power quality (PQ) monitoring campaigns.

3.2 Detection Methods Based on HOS

In future networks, changes are expected in voltage sags at most locations in the grid due to faults at the power transmission level. In addition, weather-related faults can increase the level of sags. In this sense, measures will be taken to avoid these events, such as moves for sensitive customers to secure islanding modes during such a period. The distributed generation will impact shallow sags (faults upstream of the generator and the customer downstream of the generator) [76, 77] which demand more coordination and faster fault clearing. Additionally, slower fault clearing will cause longer sags, which can be symmetrical or non-symmetrical, such as in the case of wind turbine applications [78].

Power Quality Measurement and Analysis Using Higher-Order Statistics: Understanding HOS Contribution on the Smart(er) Grid, First Edition. Olivia Florencias-Oliveros, Juan-José González-de-la-Rosa, José-María Sierra-Fernández, Manuel-Jesús Espinosa-Gavira, Agustín Agüera-Pérez, and José-Carlos Palomares-Salas.

The immunity of the equipment due to voltage sags requires further research. Indeed, a better understanding of the characteristics of voltage sags is needed. In addition, there is a demand for standards methods that calculate additional characteristics of voltage dips, including phase-angle jumps.

Related to frequency variations, a large increase in wind and solar power is expected to result in an increase in frequency variations, which should be monitored. See reference [79] for information on microgrids. Additionally, in microgrid systems, voltage and frequency variations create adverse operating conditions for load operations. Harmonics will increase as a result of the new loads and the mixed effect of new devices in the network [20]. The manifestation of solar activity (flares, bursts and others) occurs over the whole Sun, but as most radio astronomy observations are made from the Earth's surface, a significant part of solar radio events (those from the far side of the Sun) is not available for terrestrial observers.

3.3 Experiment Description

- Symmetrical and sinusoidal. A dataset of sags and transients were designed virtually in MATLAB™ in order to extract the sag disturbance regions within the HOS planes. The inputs of the experiment were to establish different criterion that HOS extracts from waveform distributions.
- Symmetrical and non-sinusoidal.
- Non-symmetrical and non-sinusoidal.

3.3.1 Computational Strategy

The strategy of computation is based on the translation from the time domain to the HOS planes. The original data sets come from the time domain, in which the analysis window length is set to 0.02 seconds (one cycle of the power-line signal). Data have been sampled at 20 kHz (400 samples per cycle).

By sweeping the sliding window (without overlapping) through the register, for each cycle the values of the three former estimators are computed. Thus, a steady-state triplet characterises the state of the network. Each cycle in the time domain is associated with a pair in each bi-dimensional HOS plane. Then, as the analysis window moves through the data record, the trajectories are depicted in three different 2D planes: variance vs. skewness, variance vs. kurtosis and kurtosis vs. skewness. While the signal waveform is unaltered, the triplet is constant and the trajectories in the planes are reduced to a single point. However, if the sliding window bumps into a disturbance, this triplet changes, and a trajectory is depicted as the point moves within the 2D planes.

3.3.2 HOS for Sag Detection under Symmetrical and Sinusoidal Conditions

According to electromagnetic compatibility (EMC), Parts 2–8: Environment, voltage dips and short interruptions on public electric power supply systems with statistical measurement results (IEC/TR 61000-2-8:2002), sags are characterised by their reference voltage and the event duration. Using the trajectory method, the reference voltage is transferred to the statistics domain, extracting the relevant information to establish compliance regions in the HOS planes. The values are summarised in Table 3.1. The variance is the statistic that detects amplitude changes (see Figure 3.1).

Table 3.1 HOS for sag detection under voltage reference in symmetrical and sinusoidal conditions.

Voltage supply variations	HOS		
	v	s	k
U_{din} 1.1	0.55	0	0
Normalised U_{c}	**0.5**	**−4.1518e−16**	**1.5**
U_{din} 0.9	0.45	0	0
U_{din} 0.8	0.3614	0	0
U_{din} 0.7	0.2451	0	0
U_{din} 0.6	0.1801	0	0
U_{din} 0.5	0.1251	0	0
U_{din} 0.4	0.0800	0	0
U_{din} 0.3	0.0450	0	0
U_{din} 0.2	0.0200	0	0
U_{din} 0.1	0.0050	0	0

Source: Olivia Florencias-Oliveros taken from reference [69]. Please, notice that s and k remain in their ideal values (0 represent that there are not significant change in s and k statistics).

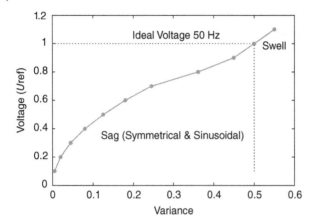

Figure 3.1 The variance for sag detection under a voltage reference in symmetrical and sinusoidal conditions. *Source:* Olivia Florencias-Oliveros taken from reference [69].

3.3.3 HOS for Sag Detection Including Phase-Angle Jump Based on Non-symmetrical and Non-sinusoidal Conditions

The **variance** helps in sag detection according to changes in the amplitude. In a sag that begins and ends in zero degrees without a phase-angle jump, skewness and kurtosis will remain at their ideal values, as shown in Table 3.1.

Nevertheless, this is not real and different phase-angle jumps have been studied in previous simulations in Chapter 2 when cycle-to-cycle amplitude changes the skewness and kurtosis detected those phase-angle jump deviations as a result of transition states in the evolution (Figures 2.3 and 2.4).

The **skewness** helps in detection of the events because it finds the phase-angle jumps that come with a non-symmetrical and non-sinusoidal behaviour between different states when an event has taken place (see skewness definition in Chapter 2, Section 2.4.1). In addition, it helps to clarify whether the event is symmetrical or not (see Figure 3.2, left).

The **kurtosis** helps in detection of the events because it finds the phase-angle jumps that come with a non-sinusoidal behaviour. The kurtosis computations equal a symmetrical event rather than a non-symmetrical one (see Figure 3.2, right).

Changes on the ideal voltage supply, alters the statistic values cycle by cycle, detecting those transient events and permanent events through HOS. Experiments simulated in lab can be applied and interpret on line-to neutral measurements. Following the previously criterion, the family of curves obtained show the dependency ratio of the coefficients (variance, skewness and kurtosis) to the residual voltage for sinusoidal

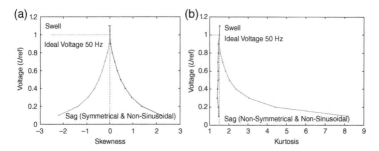

Figure 3.2 HOS behaviour vs. changes in the reference voltage (*U*ref) according to sags and swell detection. (a) Evolution of skewness vs *U*ref. (b) Evolution of kurtosis vs *U*ref. *Source:* Olivia Florencias-Oliveros taken from reference [69].

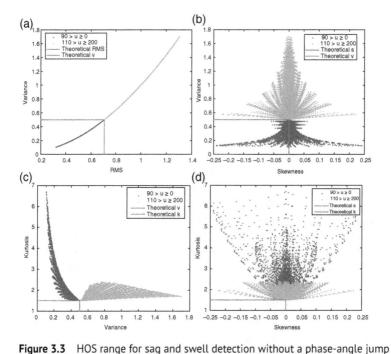

Figure 3.3 HOS range for sag and swell detection without a phase-angle jump under non-symmetrical and non-sinusoidal conditions, changes on one cycle. (a) RMS vs variance. (b) Skewness vs variance. (c) Variance vs kurtosis. (d) Skewness vs kurtosis. *Source:* Olivia Florencias-Oliveros taken from reference [69].

signals, that exhibit changes on 1 cycle and changes on 10 cycles, simulations with 16 phase-angle jumps under non-symmetrical and non-sinusoidal conditions are introduced in Section 3.3.3.1.

All the measurements have been represented through the HOS planes shown in Figure 3.3 and are summarised in Tables 3.2–3.4. Appendix C,

Table 3.2 HOS for sag detection without phase-angle jumps under voltage reference in non-symmetrical and non-sinusoidal conditions.

Voltage supply variations	HOS		
	v	s	k
Normalised U_c	**0.5**	**−4.1518e−16**	**1.5**
U_{din} 0.9	0.42–0.5	−0.008–(0.008)	1.5–1.52
U_{din} 0.8	0.35–0.5	−0.02–(0.02)	1.5–1.58
U_{din} 0.7	0.3–0.5	−0.03–(0.03)	1.5–1.7
U_{din} 0.6	0.25–0.5	−0.06–(0.06)	1.5–1.95
U_{din} 0.5	0.2–0.5	−0.08–(0.08)	1.5–2.4
U_{din} 0.4	0.15–0.5	−0.1–(0.1)	1.5–3
U_{din} 0.3	0.15–0.45	−0.15–(0.15)	1.5–4
U_{din} 0.2	0.1–0.45	−0.2–(0.2)	1.5–5.5
U_{din} 0.1	0.1–0.45	−0.25–(0.25)	1.5–7

Source: Olivia Florencias-Oliveros taken from reference [69].

Table 3.3 HOS for swell detection without phase-angle jumps under non-symmetrical and non-sinusoidal conditions.

Voltage supply variations	HOS		
	v	s	k
Normalised U_c	**0.5**	**−4.1518e−16**	**1.5**
U_{din} 1.1	0.5–0.545	−0.003–(0.003)	1.499–1.506
U_{din} 1.2	0.5–0.62	−0.015–(0.015)	1.5–1.535
U_{din} 1.3	0.5–0.75	−0.02–(0.02)	1.51–1.58
U_{din} 1.4	0.5–0.85	−0.04–(0.04)	1.54–1.66
U_{din} 1.5	0.5–0.95	−0.06–(0.06)	1.54–1.74
U_{din} 1.6	0.5–1.1	−0.08–(0.08)	1.6–1.85
U_{din} 1.7	0.5–1.2	−0.1–(0.1)	1.6–1.95
U_{din} 1.8	0.5–1.3	−0.15–(0.15)	1.65–2.05
U_{din} 1.9	0.5–1.5	−0.17–(0.17)	1.65–2.2
U_{din} 2	0.5–1.7	−0.22–(0.22)	1.65–2.35

Source: Olivia Florencias-Oliveros taken from reference [69].

Table 3.4 HOS range for sag detection including phase-angle jump based on non-symmetrical and non-sinusoidal conditions in a time-series of 10 cycles.

Voltage supply	HOS		
	v	s	k
U_{din} 1.1	0.55	0.03	1.512
Normalised U_c	**0.5**	**−4.1518e−16**	**1.5**
U_{din} 0.9	0.45	−0.03	1.49
U_{din} 0.8	0.3614	(−0.1021)−0.1021	1.4493−1.5482
U_{din} 0.7	0.2451	(−0.2182)−0.2182	1.4339−1.6718
U_{din} 0.6	0.1801	(−0.3288)−0.3288	1.4399−1.8255
U_{din} 0.5	0.1251	(−0.4729)−0.4729	1.4358−2.0301
U_{din} 0.4	0.0800	(−0.6681)−0.6681	1.4100−2.3693
U_{din} 0.3	0.0450	(−0.9470)−0.9470	1.4002−3.0472
U_{din} 0.2	0.0200	(−1.3801)−1.3801	1.4520−4.4293
U_{din} 0.1	0.0050	(−2.1975)−2.1975	1.4685−8.1530

Source: Olivia Florencias-Oliveros taken from reference [69].

HOS Range for Sag Detection in 1 cycle, and Appendix D, HOS Range for Sag Detection in 10 cycles summarise both of the different HOS trajectories that follow sag and swell in different residual voltages.

Figure 3.3 defines the HOS range for sag and swell detection without phase-angle jumps under non-symmetrical and non-sinusoidal conditions when the events take place in 1 cycle. Figure 3.4 and Table 3.2 clarified the different regions for sag. Figure 3.5 and Table 3.3 characterise swell, respectively, according to the voltage reference magnitudes in Figure 3.3.

Tables 3.2 and 3.3 summarise both the different HOS ranges according to the voltages reference changes in sag and swell events under non-symmetrical and non-sinusoidal conditions, as given in Figure 3.3.

In Figure 3.4 the different regions related to sag in the HOS planes are identified with voltage reference changes and amplitude changes. Notice that kurtosis and skewness follow the arc shapes that have been described in Figure 3.2. The evolutions of the different amplitude reference regions are summarised individually in Appendix C, HOS Range for Sag Detection in 1 cycle.

Figure 3.5 shows swell regions in the HOS planes that are identified by a voltage reference increase and amplitude changes. Notice that kurtosis and skewness detect the different magnitude ranges.

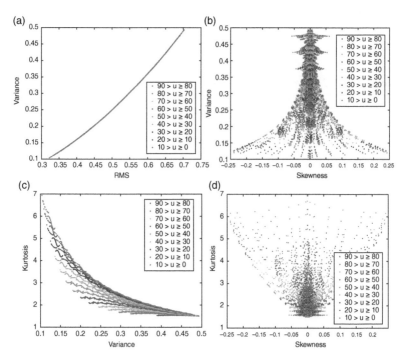

Figure 3.4 HOS range for sag detection in Figure 3.3 without a phase-angle jump under non-symmetrical and non-sinusoidal conditions. (a) RMS vs variance. (b) Skewness vs variance. (c) Variance vs kurtosis. (d) Skewness vs kurtosis. *Source:* Olivia Florencias-Oliveros taken from reference [69].

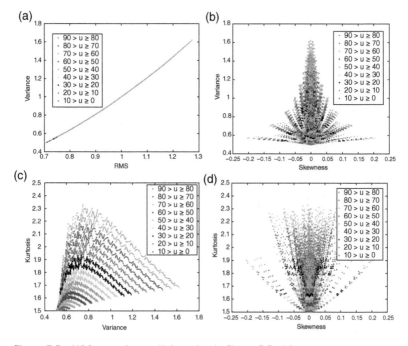

Figure 3.5 HOS range for swell detection in Figure 3.3 without a phase-angle jump under non-symmetrical and non-sinusoidal conditions. (a) RMS vs variance. (b) Skewness vs variance. (c) Variance vs kurtosis. (d) Skewness vs kurtosis. *Source:* Olivia Florencias-Oliveros taken from reference [69].

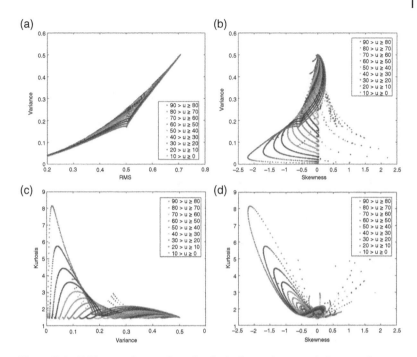

Figure 3.6 HOS range for sag detection including a phase-angle jump under non-symmetrical and non-sinusoidal conditions, in a time series of 10 cycles. (a) RMS vs variance. (b) Skewness vs variance. (c) Variance vs kurtosis. (d) Skewness vs kurtosis. *Source:* Olivia Florencias-Oliveros taken from reference [69].

Table 3.4 and Figure 3.6 show the sag evolution when 10 cycles are taken into account. Notice that the trajectories follow new transition states. The evolutions of the different amplitude reference regions are summarised individually in Appendix D, HOS Range for Sag Detection in 10 cycles. Finally, different regions for sag events are clarified in Figures 3.7–3.9.

After the simulation, the variance, skewness and kurtosis range have been defined. The outside regions of the trajectories in Figures 3.7–3.9 have a greater probability of presenting other kinds of events, such as transients, which are not related to sag.

The 2D HOS plane in previous figures with sag thresholds helps to improve events detection in virtual instruments that incorporate statistics. The method was previously introduced in reference [74], where the 2D patterns univocally defined the electrical perturbation.

The method is easy to implement using a hand-held PQ instrument and discriminates transients according to their sources. The 2D HOS graphs reveal the evolution of events.

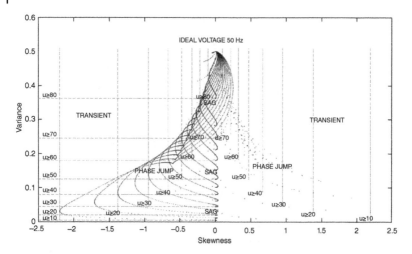

Figure 3.7 Skewness vs. variance, sags and phase-angle jumps regions. *Source:* Olivia Florencias-Oliveros taken from reference [69].

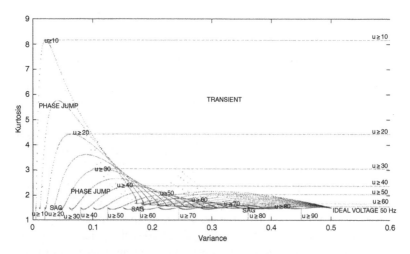

Figure 3.8 Variance vs. kurtosis, sags and phase-angle jumps regions. *Source:* Olivia Florencias-Oliveros taken from reference [69].

Events detected in the field are analysed in Figures 3.10–3.12, showing the sag and transients. They were acquired at 20 kHz (400 samples per cycle) in a building measurement node.

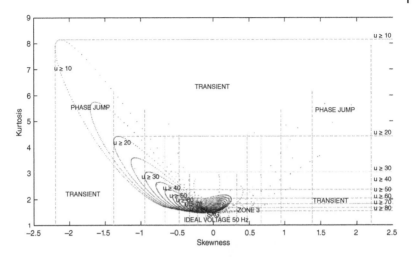

Figure 3.9 Skewness vs. kurtosis, sags and phase-angle jumps regions. *Source:* Olivia Florencias-Oliveros taken from reference [69].

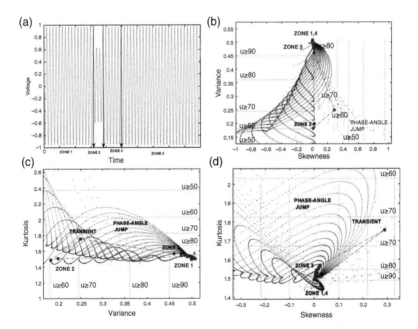

Figure 3.10 Sag 60% event in the HOS planes with the phase-angle jumps. (a) Sag event in the time-domain. (b) Sag trajectory represented on the skewness vs. variance plane, notice skewness detect phase-angle jump. (c) Sag trajectory represented on the variance vs. kurtosis plane. (d) Sag trajectory represented on the skewness vs kurtosis plane, notice the transition state detection into the sag evolution. *Source:* Olivia Florencias-Oliveros taken from reference [69].

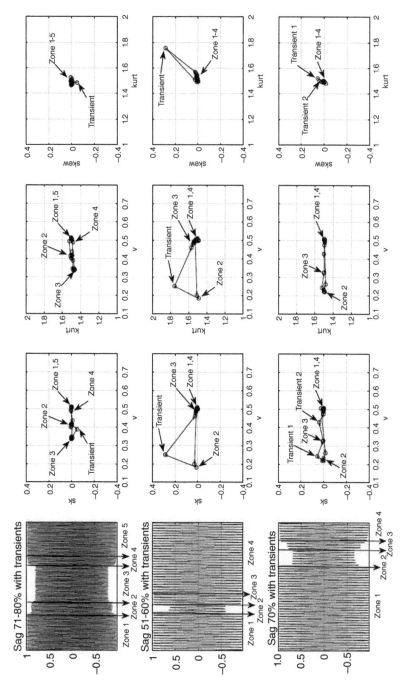

Figure 3.11 In field measurements that contain coupled transients resulting from the voltage gaps. *Source:* Olivia Florencias et al. taken from reference [74].

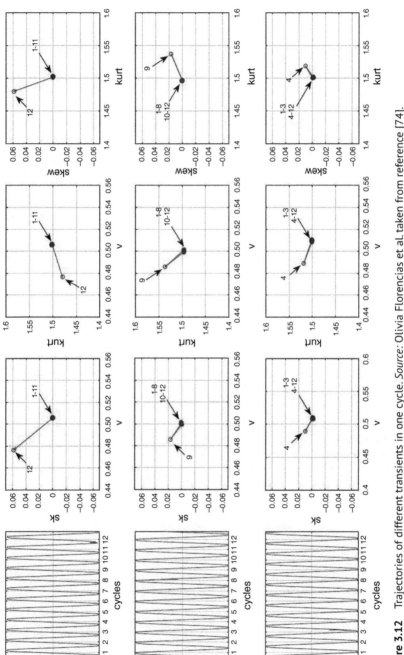

Figure 3.12 Trajectories of different transients in one cycle. *Source:* Olivia Florencias et al. taken from reference [74].

3.3.3.1 HOS Range for Transient Detection Including Phase-Angle Jump Based on Non-symmetrical and Non-sinusoidal Conditions

Table 3.5 helps to settle the different HOS regions aligned to transient detection and including the phase-angle jumps under non-symmetrical and non-sinusoidal conditions.

The different criteria on which the analysis is based allow both events already described to be studied, capturing those of another nature for later analysis. The flow diagram in Figure 3.13 helps the proposed strategy to be introduced as part of a virtual instrument that performs an off-line analysis of a data set or an online analysis aimed at detecting events.

Changes in the amplitude are indicative of different events such as inter-ruption (<0.1 pu), sag (0.1–0.9 pu) and swell (1.1–1.8 pu). Additionally, event duration helps to categorise more deeply the short duration, such as the impulsive transients, instantaneous events between 0.5 and 30 cycles, momentary 30 cycles–3 seconds and temporary between >3 seconds and 1 minute.

The skewness and kurtosis remain at their ideal values (then the signal is sinusoidal or the sag agrees with the sliding window). Skewness and kurto-sis change their ideal values, so the signal is non-sinusoidal. If the skewness values agree in positive and negative values, then the event is symmetrical.

Table 3.5 HOS range for transient detection including phase-angle jump based on non-symmetrical and non-sinusoidal conditions.

| Voltage supply variations | HOS | | |
	v	s	k
Normalised U_c	**0.5**	**−4.1518e-16**	**1.5**
U_{din} 0.9	0.45	−0.03	1.49
U_{din} 0.8	0.3614	([−0.014]−0.014)	1.5–1.55
U_{din} 0.7	0.2960	([−0.2]−0.2)	1.5138–1.7161
U_{din} 0.6	0.2440	([−0.0552]−0.0552)	1.5245–1.9532
U_{din} 0.5	0.2	([−0.0859]−0.0859)	1.5365–2.3793
U_{din} 0.4	0.1640	([−0.1252]−0.1252)	1.5485–3.0744
U_{din} 0.3	0.1360	([−0.1706]−0.1706)	1.5596–4.1870
U_{din} 0.2	0.1160	([−0.2137]−0.2137)	1.5688–5.6108
U_{din} 0.1	0.1040	([−0.2399]−0.2399)	1.5774–6.9393

Source: Olivia Florencias-Oliveros taken from reference [69].

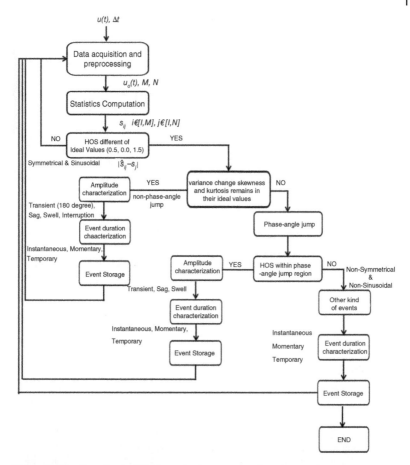

Figure 3.13 Flowchart of HOS monitoring strategy focusing on detecting short-duration events based on the waveform changes. *Source:* Olivia Florencias-Oliveros taken from reference [69].

3.4 Flow Diagram of HOS Monitoring Strategy Focus on Detecting Short Duration Events: Detecting Amplitude, Symmetry, and Sinusoidal States

According to the IEEE Recommended Practices for Monitoring Electric Power Quality, different phenomena of power systems exhibit typical characteristics. Some of them are detected through the root-mean-square (RMS) method. Table 3.6 summarises the events and the HOS domain information aligned to different categories.

Table 3.6 Categories and typical characteristics of the power system electromagnetic phenomenon [12] that are potentially characterised through the HOS event detection method.

Categories	Typical duration	Typical voltage magnitude	RMS range	HOS range
Instantaneous				
Transients	0.1 ms rise	>1 ms	Figure 3.12	Table 3.5
Sags	0.5–30 cycles	0.1–0.9 pu	Figure 3.2 up to Figure 3.10	Tables 3.1, 3.2, 3.4
Swells	0.5–30 cycles	1.1–1.8 pu	Figures 3.3 and 3.5	Table 3.3
Momentary				
Sags	30 cycles–3 s	—	Figure 3.2 up to Figure 3.10	Tables 3.1, 3.2, 3.4
Swells	30 cycles–3 s	1.1–1.4 pu	Figures 3.3 and 3.5	Table 3.3
Temporary				
Interruption	>3 s–1 min	<0.1 pu	—	—
Sag	>3 s–1 min	0.1–0.9 pu	Figure 3.2 up to Figure 3.10	Tables 3.1, 3.2, 3.4
Swells	>3 s–1 min	1.1–1.2 pu	Figures 3.3 and 3.5	Table 3.3
Long duration				
Undervoltages	>1 min	0.8–0.9 pu	Figure 2.4 (Chapter 2)	Below values in Tables 2.6 and 2.7
Overvoltages	>1 min	1.1–1.2 pu	Figure 2.4 (Chapter 2)	Above values in Tables 2.6 and 2.7

Source: Olivia Florencias-Oliveros taken from reference [69].

Findings: The trajectory-based method reveals the conformity regions in the HOS planes according to different standardised events and their steady state, which considers the depth of the event, the duration and the phase-angle jumps between states under non-symmetrical and non-sinusoidal conditions. Nevertheless, using the HOS planes is possible to detect and classify not only single events but also multiple ones. In addition, the method helps to discriminate a transitory event from one that comes aligned to a sag event.

Recommendations: CIRED/CIGRÉ propose detection of phase-angle jumps. Most of the events that are present in the smart grid (SG) consist of different types of sags that can appear in different forms, depicted in Figures 3.1 and 3.2, and are classified according to the standard UNE-IEC/TR 61000-2-8 [80]. In this frame, our contribution to this norm consists of the establishment of new criteria to discriminate phase-angle jumps and their trajectory in the HOS planes related to sag events. The proposed method is capable of depicting a phase-angle jump classification aligned to improve sag characterisation when using HOS.

3.5 Continuous Events Characterisation Fundamental Frequency

Instantaneous frequency deviations are another case of steady-state changes under symmetric and sinusoidal power supply conditions. Some techniques such as Lissajous figures or the Bowditch curves have been previously proposed to compare two signals with different frequencies in 2D graphs. The relationship between the frequencies is observed in the trajectories (X–Y movements), which exhibit specific patterns such as circular, parabolic or elliptical types.

Frequency measurements based on HOS have been previously introduced in Chapter 2. Nevertheless, frequency range changes according to the standard are explained here. Table 3.7 summarises frequency ranges according to the standard EN 50160:2015.

The sliding window method is used to extract the HOS cycle-to-cycle and to introduce a degree of measurement uncertainty related to the fixed window. A voltage cycle analysed through the sliding window (0.02 second for 50 Hz) has more or less information (points) that depends on their fundamental frequency. The proposal method characterises fundamental frequency uncertainty in the HOS planes in order to extract that information for PQ measurements according to the standards.

Table 3.7 Industrial frequency ranges according to EN 50160:2015 [13].

Network characteristics	Sinusoidal and symmetrical waveform	Frequency range according to the standard
Ideal voltage supply	**50 Hz**	—
Synchronous connection and an interconnected system	49.9 Hz–50.1 Hz	±0.1 Hz for 100%
Non-synchronous connection to an interconnected system	48 Hz–52 Hz	±2 Hz for 100%

Source: Authors.

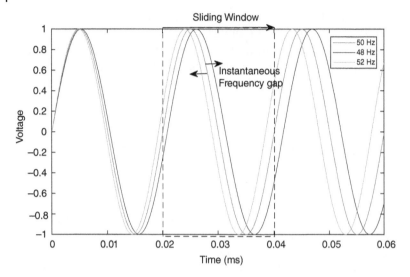

Figure 3.14 Instantaneous frequency gap that can be detected through the sliding window when a signal deviates from their fundamental frequency (50 Hz, at 20 kHz, 400 points per cycle). *Source:* Olivia Florencias-Oliveros et al. taken from reference [81].

Figure 3.14 shows the instantaneous frequency gaps that can be detected through feature extraction when use the sliding window method. Changing the period of the signal means that more or less information is computed by the window (the period is the inverse of the fundamental frequency). A waveform of 48 Hz exhibits 416 points per cycle while a waveform of 52 Hz exhibits 385 points.

For example, in Table 3.8 fundamental frequency changes on sinusoidal and symmetric signals are changes in the period, which means that a

Table 3.8 Frequency deviations on sinusoidal and symmetric signals.

Sinusoidal waveform	Info per cycle at 20 kHz	Difference from the ideal voltage supply	v	s	k
48 Hz	416	+15 points	0.52	±0.0225	1.4456
49.5 Hz	400	—	0.5023	±1.1992	1.4970
50 Hz	**400**	—	**0.5**	**0.00**	**1.5**
50.1 Hz	400	—	0.5003	±1.3448	1.5030
52 Hz	385	−15 points	0.4827	±0.019050	1.5552

Source: Olivia Florencias-Oliveros et al. taken from reference [81].

waveform with the highest (lowest) frequency exhibits a lowest (highest) period, thus showing that more or less information is computed by the sliding window.

In field measurements, the instantaneous frequency gap can be detected according to the standard. Test and measurement techniques for PQ instrumentation according to the UNE 61000-4-30 standard quantifies the fundamental frequency as the number of periods counted during a clock time interval of 10 seconds, divided by the accumulated measurement time. However, frequency changes depend on the information within each cycle. In this sense, parameters such as phase angles, zero crossings and the quantization of a whole number of cycles, as established by the standard, will inevitably be done in different measurement intervals if there is not an industrial frequency synchronisation system for multiple locations.

According to UNE 61000-4-30:2015 [1], the uncertainty of the measurement for the industrial frequency must be specified for the range of measurements from 48 to 52 Hz, not exceeding ±2 Hz, for a class A instrument. This criterion had been chosen in order to map the frequency variations onto the HOS planes.

The output resolution is capable of reporting whether the frequency of the supply signal falls instantaneously within the ranges allowed by the regulations as well as determining the frequency from the average values over 10-seconds of measurement times. These instantaneous deviations exhibit a behaviour that is translated into HOS planes. At the same time, the uncertainty of the statistic measurement is reduced as well as the uncertainty of the indices that can be developed from them.

Each signal under test is characterised by a certain trend, depicted in a trace. As a result, the proposed PQ index computes the statistical deviation. The path a statistic (x, y statistics) follows as it moves through the HOS plane is the trajectory.

A battery of test signals has been used in order to study the behaviour of the HOS within the different thresholds specified by the normative and to map the frequency shifts into the HOS plane.

3.5.1 Frequency Deviation Regions in the HOS Planes

HOS frequency uncertainty has been tested in the HOS planes. The statistics ranges related to maximum frequency deviation in synchronous networks and non-synchronous networks are summarized in Tables 3.9 and 3.10.

The following pair of figures shows the evolution of the paths (trajectories) that are followed by the statistics in the HOS planes for different

Table 3.9 HOS and PQ fluctuations for different frequency ranges: synchronous in interconnected systems.

	Ranges in the HOS planes for synchronous networks			
	Statistics			
Frequency (Hz)	**Variance**	**Skewness**	**Kurtosis**	**PQ index**
49.9	[0.5023,0.5002]	[0.004,−0.004]	[1.5015,1.4970]	[0.0065,0.0041]
50.1	[0.5022,0.5003]	[0.004,−0.004]	[1.5030,1.4985]	[0.0069,0.0033]

Source: Olivia Florencias-Oliveros et al. taken from reference [81].

Table 3.10 HOS and PQ fluctuations for different frequency ranges: non-synchronous networks in isolated systems.

	Ranges in the HOS planes for non-synchronous networks 48–52 Hz			
	Statistics			
Frequency (Hz)	**Variance**	**Variance**	**Variance**	**PQ index**
48	[0.5219,0.4790]	[0.0826,−0.0826]	[1.5391,1.4412]	[0.1253,0.0820]
52	[0.5187,0.4822]	[0.0762,−0.0763]	[1.5587,1.4787]	[0.1017,0.0779]

Source: Olivia Florencias-Oliveros et al. taken from reference [81].

frequencies, aligned with the requirements of the UNE 50160:2015. Figure 3.15 shows the evolution of the different registers (10-seconds in length) in a frequency interval of 49–51 Hz. All the trajectories are centred in the ideal values, corresponding to the true values of the statistics (variance = 0.5, skewness = 0.0, kurtosis = 1.5). It is observed that constant-frequency deviations within the studied range exhibit specific paths in the HOS plane. Variance vs. skewness paths adopt semi-circumference shapes, while the variance vs. kurtosis paths follow circumference arcs. Finally, the skewness vs. kurtosis trajectories adopt the forms of semi-ellipses. The curves in Figure 3.15 offer a calibration background over which dynamic measurements are developed and tracked

Figure 3.16 shows the calibrated HOS plane of Figure 3.15 with different trajectories of the signals. The highlighted paths correspond to the new measurements developed based on the calibrated planes, which have been coloured in light grey. It is possible to observe the superposition of a signal

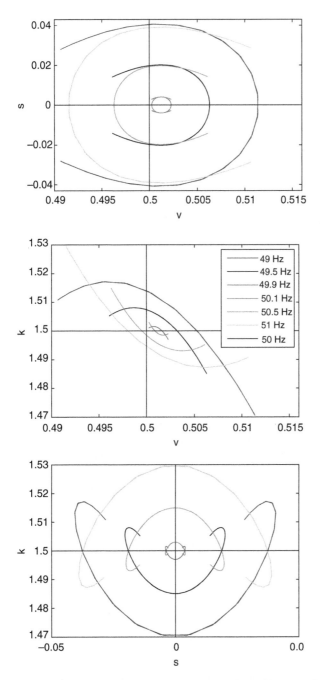

Figure 3.15 Constant frequency deviations in the HOS planes. *Source:* Olivia Florencias-Oliveros et al. taken from reference [81].

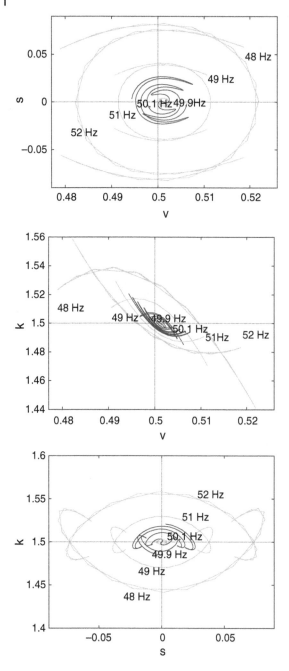

Figure 3.16 Tracking new measurements on the background screen floor. *Source:* Olivia Florencias-Oliveros et al. taken from reference [81].

that changes its frequency cycle-to-cycle in the range 49.9–50.1 Hz (synchronous networks, UNE-EN 61000-4-30:2015). The example shows how the frequency fluctuations of 48–52 Hz (isolated networks, 61000-4-30:2015) are depicted in the HOS plane. The simple fact of superimposing paths may make it possible to match the current measured trajectory with a previously calibrated one.

3.5.2 Flow Diagram of HOS Focus on Detecting Fundamental Frequency Deviations

The results may suggest potential applications in the field of PQ instrumentation, and more specifically in portable equipment that is currently deployed in the electrical network following a flowchart such as in that shown in Figure 3.17.

Findings: From a general point of view, this method has connected power system instability and frequency uncertainty, in order to detect PQ deviations due to frequency uncertainly in the HOS planes. It has been shown that deviations from the frequency can be studied from the perspective of a bi-dimensional plane, considering the intervals currently used by the UNE 61000-4-30 standard. Thus, the study of the industrial frequency through trajectories could be a tool for improving the visual representation of the quality of the energy supplied, providing information beyond traditional average values. The developed tool incorporates the calibration of the zone of deviations of the frequency (allowed in the norm EN 50160:2015), information on the direction of the trajectories (potentially used in future norms), as well as the influence of statistical deviations on the developed PQ indices.

More specifically, three main ideas have been proposed:

- First, a translation from the fundamental frequency to the HOS domain has been performed in order to elicit how frequency deviations can be measured in the HOS planes.
- Second, it is confirmed that the signals trajectories in three statistics vs. statistic planes have demonstrated that each electrical disturbance exhibits a different and characteristic path, showing that frequency deviations affect PQ in dynamic measurements. The calibrated background settles down a pattern over which new measurements (dynamic paths) are allocated with the goal of estimating the frequency of the measured signal over a 10-second period. The frequency patterns in kurtosis vs. skewness allow trajectories of frequency deviations similar to the Lissajous curves to be obtained. This similarity illustrates the capability of the HOS planes

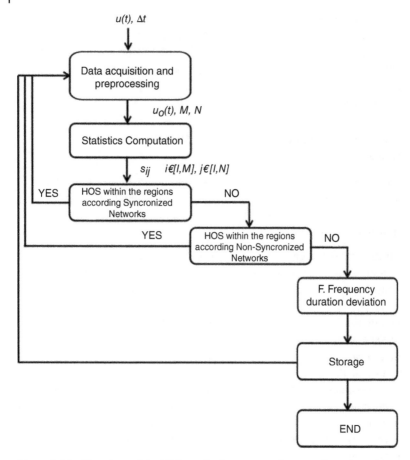

Figure 3.17 Flowchart of the HOS monitoring strategy focus on detecting fundamental frequency deviations based on the waveform changes. *Source:* Olivia Florencias-Oliveros taken from reference [69].

to discriminate not only transient disturbances but also continuous PQ phenomenon, like frequency fluctuations. In this sense, more real-life signals with frequency deviations would show that all the capabilities of the method are currently under research.

- Third, a global PQ index based on differential statistic measurements accompanies the 2D HOS traces and indicates how far the shape of the delivered voltage is from their ideal nominal pattern. This idea may also be optimised using machine learning techniques, along with the automatic establishment of the weights in the PQ index.

Recommendations: As a final thought, the results are of potential interest for the synchronised calculation of the power system frequency in PQ applications using phasor measurement units (PMUs). This may be done by synchronising PMUs via a global positioning system (GPS) and using the time protocols included in the IEEE standard 1588:2008 for the Precision Clock Synchronisation Protocol for Networked Measurement and Control Systems [82].

In addition, the method is an alternative to test the fundamental frequency in a wind turbine when connecting it to the network.

In a continuous monitoring strategy based on HOS (see Chapter 5), the fundamental frequency uncertainty obtained in this section would be quantified as well as their maximum deviation. A more deviated steady-state waveform has a different disturbance origin.

3.6 Detection of Harmonics with HOS in the Time Domain

New electronic devices connected to power systems are a source of multiple non-linear loads that pollutes the network emitting harmonics and inter-harmonics. The harmonics characterisation and emission control is ruled by international standards, mainly IEEE standards and those of the IEC (International Electrotechnical Commission), but each country has their own normative. Nevertheless, reference levels are well established with the objective of endowing companies with surveillance tools to deploy in their networks.

The IEC standards have been adopted by the European Union (EU) where CENELEC has been adopted in different categories, such as part 1 (Generalities) to part 6 (Several). The standards relatives to harmonics are in the second category (Ambient-IEC 61000-2-xx) and third category (Emission limits-IEC 61000-3-xx). In the fourth category (Test and measurement techniques-IEC 61000-4-xx) are harmonics and inter-harmonics.

The UNE-EN 61000-2-2 defines the compatibility levels for disturbances conducted by low frequency in LV public systems such 50 and 60 Hz one-phase or three-phase systems with nominal tension until 240 and 415 V. Compatibility levels indicated are intended to reduce the number of malfunction complaints to an acceptable level. In some special cases, these levels can be exceeded.

EMC is 'the ability of any device, equipment or system to function satisfactorily in its electromagnetic environment without causing electromagnetic disturbances in that environment'. Relative to PQ and EMC issues, two questions are mainly identified; the first concerns the consumer equipment (load) connected to the distribution system and the second concerns the distribution and transmission systems. Harmonics emission issues caused by power electronics devices-based converters and DC technologies applied recently in the grid are a particular research field within the PQ issues.

In the next two sections, an HOS approach to infer harmonic distortion from the time-domain measurements is introduced, according to the voltage harmonic limits in the PCC of an installation. HOS behaviour under harmonics distortion from the time domain is summarised in the conclusions section.

3.6.1 Time Domain Analysis

Time-domain information helps to extract the peak factor indicator, which is the easiest technical way to detect harmonics. It is a measure that relates the waveform peak value with the true effective value (RMS). As discussed in Chapter 2, an ideal sinusoidal unit present is by definition a crest factor of 1.414. When the current is distorted, the peak factor achieves values over or below 1.41, which could be interpreted as a non-linear effect on the analysed circuit.

Therefore, any deviation of 1.414 could be interpreted as a distorted waveform. Despite that, this technique is not strictly rigorous because it does not consider harmonic frequencies in magnitude or in phase (information from the frequency domain). Some common peak values are 1.414 for lineal loads (without harmonics), between 2 and 3 when measuring computer equipment (the source of harmonics) and 2 for speed regulators (the source of harmonics).

In addition, total harmonic distortion (THD) is the harmonic distortion present in the measurement point. The THD is the relationship between the efficient value of harmonic components (voltage or current) and the efficient value of the fundamental components (voltage or current). It is a deviation measure with respect to the sinusoidal waveform and can be defined by both current distortion (THDi) (Eq. 3.1) and voltage distortion (THDi) (Eq. 3.2):

$$\text{THD} = \frac{\sqrt{\sum_{i=2}^{\infty}\left(I_i\right)^2}}{I_1}\,100\% \tag{3.1}$$

$$\text{THD} = \frac{\sqrt{\sum_{i=2}^{\infty}\left(V_i\right)^2}}{V_1}\,100\% \tag{3.2}$$

While h denotes the harmonic order, 1 is the fundamental and M is the voltage (V) or the current (I). The more distorted a signal is, the greater will be the harmonic distortion factor, and as a consequence the presence of harmonics in the signal will increase. Similar to the peak factor, the THD gives the effective values of harmonics in the waveform, but does not provide any information on harmonic phase angles.

In addition, when calculating THD, all harmonics are examined equally, so there is a loss of information relative to the harmonic order and magnitude. The THD does not distinguish between two signals with the same number of harmonics, with equal magnitude but different order or two signals completely different from the point of view of the harmonic content and the phase angle.

From the waveform quality perspective, voltage and current harmonic phase angles are not relevant. However, the phase angle allows the relationship between the non-linear loads to be evaluated and the disturbance source is known.

Nevertheless, when limiting harmonic current injections by users, this does not result in an improvement in voltage distortion in power systems.

In this section, the criterion related to HOS has been developed to quantify the degree of deviation of the waveform, although the loads themselves are not characterised and the focus is on the whole a characterisation of its effects on voltage. Here, explicitly, the impact of voltage harmonics in the time-domain is studied.

Is known that, when harmonics are present in the network the waveform is distorted. According to the level of emissions between the devices and the network, the harmonic distortion can be higher or lower and its voltage may be affected.

3.6.2 Preliminary Evaluation of Harmonics Through HOS

HOS from the time-domain measurements pretends to be a preliminary evaluation tool to detect early stages during measurement campaigns. Before installing harmonic-measuring equipment, an easy task would help detect whether harmonics are present. Based on the simulations, some simple criterion to detect the presence of harmonics from the HOS feature extraction analysis can be added to the others proposed by IEEE 1159-2014 [83]:

- Is the crest factor of voltage or current different from 1.4?
- Is the factor form different from 1.1?
- Are true RMS measurements different from 0.7?
- Do kurtosis below 1.43 and variance and skewness remain in their theoretical values of 0.5 and 0.0, respectively?

This criterion helps to decide whether to trigger a deeper analysis with the harmonic monitoring objective. A flow diagram helps to introduce an HOS measurement procedure for harmonics detection in Figure 3.18.

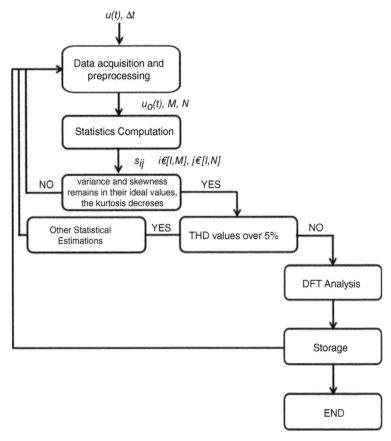

Figure 3.18 Flow diagram that helps introduce HOS measurements as a harmonic detection method. *Source:* Olivia Florencias-Oliveros taken from reference [69].

Findings: One of the advantages of using HOS through this method is that it is possible to discriminate between a transient non-sinusoidal condition and a harmonic distortion over time. If it is a transient state, variance and skewness would change their values, but under a harmonic steady state variance and skewness retain their ideal values while kurtosis would change to detecting a non-sinusoidal condition.

Recommendations: It is difficult from the HOS perspective to define directly what type of harmonics affects the monitored node. However, harmonic deviations can be detected through kurtosis deviation and both stable values of variance and skewness. The method would help to trigger further specific analyses of other methods used in the frequency domain, such as DFT and the previously analysed method in Chapter 1, Section 1.4.8.

3.7 Conclusions

HOS-based techniques were proposed in order to detect different events from the time-domain method that have not been characterised until now in the HOS domain: sag phase angle jumps, swells, harmonics and fundamental frequency.

In addition, symmetrical and non-symmetrical voltage sags considering phase angle jumps are useful in different applications.

The fundamental frequency instantaneous deviations impact HOS measurements in the bi-dimensional plane. The uncertainty detected in the HOS planes helps to correct further models based on these statistics. This is especially relevant because they must be included in future instrumentation solutions based on HOS.

These individual analyses help us to understand the different combinations of indices according to the characteristics of events in the network.

4

Measurements in the Frequency Domain

4.1 Introduction

In this chapter, a completely different vision of the power signal is given in order to detect different disturbances. This is a frequency-domain analysis, where a signal segment is decomposed into its frequency components (sinusoidal signals of different frequencies that create it). Once decomposition is done, higher-order statistics (HOS) is applied, extracting features. With this method, detected characteristics are not now related to the wave, but to each of the frequencies that compose it. On the other hand, a segment of the signal is needed that is much longer than in the time domain.

The most similar analysis in power systems is regulated by IEC 61000-4-7 [84], where harmonic distortion is studied. This is related to the second-order spectral (frequency-domain) analysis. However, this work is centred on the fourth-order spectral analysis, which is done using spectral kurtosis (SK), an indirect estimator based on the discrete Fourier transform (DFT). The second-order spectral analysis is blind to amplitude evolution of each frequency component (an individual frequency inside the time series), but fourth-order analysis detects and enhances amplitude evolutions, fast changes, and others variations for each frequency component.

A perfect power signal is a perfect sinusoid of 50 Hz, which is seen in the frequency domain as amplitude only for 50 Hz. Any other amplitude in other frequencies (over the random noise present in all of them) implies a waveform distortion. SK detects frequency component amplitude evolution, returning in this situation to −1 for 50 Hz, indicating that its amplitude is constant all the time, with a near-to-zero value for only noise frequency components.

Power Quality Measurement and Analysis Using Higher-Order Statistics: Understanding HOS Contribution on the Smart(er) Grid, First Edition. Olivia Florencias-Oliveros, Juan-José González-de-la-Rosa, José-María Sierra-Fernández, Manuel-Jesús Espinosa-Gavira, Agustín Agüera-Pérez, and José-Carlos Palomares-Salas.
© 2023 John Wiley & Sons Ltd. Published 2023 by John Wiley & Sons Ltd.

Permanent waveform distortion will show as a constant amplitude, so will show the SK −1 value. Disturbances appear and disappear suddenly, so frequencies show sudden changes of amplitude. SK indicates this with a high value.

For all this, SK uses more than one DFT in the analysis, so the same frequency axis must be used. The general procedure in reference [84] implies a continuous frequency sampling fix in order to ensure power network frequency join a DFT bin (each value associated with a frequency in the DFT is called a bin). This implies a frequency axis related to the fundamental frequency, changing in each cycle. This is not allowed by SK, where the sampling rate must be constant during all of the analysis (for all DFTs considered). In this situation, as indicated in regulation, the Hanning spectral window must be used when sampling frequency is not synced with power system frequency.

The SK can be implemented in computation software or in FGPA, as stated in reference [56]. All this advances the application of SK to power quality (PQ), as can be seen in research works [56, 60, 85–89].

However, SK is not exclusive for PQ as this technique has been widely used for machine vibration analysis for a rolling bearing [90–92] or gearbox [93–95] faults. Its properties in enhancing non-Gaussian processes have been used for detection of subterranean termites [86, 96–98]. Projects concerning the detection of radio-frequency interference in microwave radiometers [99] have been developed. Correlations between electroglottographic measures and pronunciation have also been studied [100]. Apart from work on signal analysis, there have also been projects such as optimised modal analysis [101], blind separation of vibration components [102] and characterising non-stationary signals [103].

4.2 Frequency Domain

Spectral analysis or analysis in the frequency domain changes completely the way to understand a time series. While the time domain represents evolutions in each instant for any property (analysing a point or a segment), the frequency domain analyses a signal segment and returns information for each frequency of the signal analysed.

As indicated by the Fourier series, a signal can be expressed by the sum of sinusoids of different frequencies, amplitudes and initial phases. Frequencies start in a base frequency and others are integer multiples of it (in addition to zero frequency, which is a constant value). Therefore, frequencies that are obtained in the frequency domain would be multiples of a certain frequency. This implies that not all frequencies are analysed, only a set of frequencies values that are equi-distanced from them in terms of frequency. The initial frequency and the distance between frequencies are called 'frequency resolutions'.

Change from the time domain to the frequency-domain is done using the Fourier transform, which is a mathematical approximation that is hard to calculate. For time series, the DFT is used and to calculate the DFT, a specific algorithm, the fast Fourier transform (FFT), was use (due to the fact that it uses very few calculations).

Maximal frequency returned by the DFT is half of the sampling rate of the input signal (the Nyquist limit of the input signal) and the frequency resolution (df) is related to the analysed segment length $df = 1/T$ with T segment lengths in seconds. It is found that, as more time is considered for DFT analysis, more frequencies can be observed.

A frequency domain analysis returns information by each frequency analysed. In an amplitude representation of the DFT, each value associated with each frequency is called a bin. Only those that have been exact frequency analysed are considered (start at 0 and separated at the df). If any other frequencies (among the two considered) show an amplitude over the noise level, it is not ignored as, 'spectral leakage' occurs. This effect implies an appearance of amplitude in surrounding frequencies. Many frequencies are affected, even with lower amplitudes, but it can be controlled using Spectral Windows. This procedure implies an amplitude modulation in a time domain, with a previously applied DFT and a limit Spectral Leakage (signal amplitude leaks to surroundings bins) to a few surrounding bins.

In power systems, DFT is used to analyse and measure harmonics (coupled signals to a main one, with their own frequency a multiple of the Power System Frequency). T is limited by the sampling rate and can be incremented or reduced only in one sampling period. However, regulations indicated 10/12 cycles at 50/60 Hz. Under ideal conditions this is 200 ms, but power system frequency changes, and to obtain 10/12 complete cycles, one sampling period more or less is not enough. By changing the sampling frequency, the segment length can be fixed perfectly for 10/12 cycles of an actual power system frequency in 50/60 Hz networks. Moreover, having a Power System Frequency centred in one bin, all its harmonics are also centered in bins, due to the fact that they are in frequencies that are multiples of the first one. This has the additional advantage of eliminating spectral leakage of fundamental and harmonic frequencies.

However, the frequency axis is now expressed in relative frequencies to actual power system ones, and changes in any analysis cycle. For SK, some DFT data are required, and the only way to evaluate the evolution of a spectral component is to evaluate the information related to the same frequency. Therefore, a constant sampling frequency is required. It must be indicated that, according to reference [84], this implies a non-sync situation of measures, so the spectral window must be applied in order to reduce spectral leakage, in particular the Hanning window.

4.3 HOS in the Frequency Domain

DFT is a great tool to use to study signals, due to the fact that it gives information about the frequency composition of a signal. However, the information obtained is an average of the values in the analysed time. Amplitude or power can be obtained using DFT, and is the averaged amplitude or power of each frequency component over the analysed time.

This is the limitation of the second order. HOS measures other parameters, in particular it is a fourth-order measure of the tailedness of statistical distributions. It is also a measure of the ratio of points that are in the tails (far from the mean value) or a measure of how concentrated points are around the mean value.

A frequency component is a sinusoidal signal with frequency equal to the frequency analysed in the bin. For a constant amplitude, the same point distribution is obtained for all cycles. However, when there are amplitude changes and point distribution changes, the amplitude changes are due to random noise, sudden appearances, huge increases, huge decreases, amplitude oscillations, etc. In all these situations, different point distributions are observed and different point concentrations around the mean value are observed. Kurtosis is a relative measure, so no matter what the frequency component amplitude is, information is obtained related to its changes, or amplitude variability.

SK is a great tool used to determinate the duration of the appearance of each frequency component through its amplitude variability. The noise frequency component has a characteristic value (around 0), a constant amplitude frequency component (around -1), high amplitude variation frequency components and another SK value (M, which is the number of realisations or DFT used for SK calculations). These values are not absolutes, so intermediate values indicate intermediate situations. In the following sections, all these situations will be explained in detail.

The fourth-order frequency domain was first applied making use of cross-frequency correlation, generating a complex calculation process to assess nonlinear, non-stationary and non-Gaussian processes. This process is called 'Trispectrum', as used in references [104] and [105] for detecting bearing faults. Later, an estimator, called SK, for the fourth-order frequency domain was developed, based on DFT, as stated in reference [106]. It makes the calculation of fourth-order spectra simpler and easier to find.

The idea of non-Gaussian behaviour in spectral components was discovered in underwater ice cracks sounds during the FRAM II Artic expedition in 1981, as stated in reference [107]. Later, the fourth-order frequency domain was applied to many fields, such as non-stationary vibration analysis in machinery [60, 90–94], plague detection [86, 96–98], radio-frequency interference in microwave radiometers [99], correlations between electroglottographic measures and pronunciation [100], as well as work on signal analysis, such as optimised modal analysis [101], blind separation of vibration components [102] or characterisation of non-stationary signals [103].

4.3.1 Spectral Kurtosis in Power Quality

The ideal power system waveform is a perfect sinusoidal example of 50/60 Hz, which is a frequency component of a frequency decomposition of a DFT. Any distortion or disturbance of that waveform appears as another frequency component, which can appear only for an instant (for example, a cycle), for a period (for example, from a few cycles to a few seconds) or for a long time (for example, from 30 minutes to always).

SK detection capabilities, described in previous sections, fit perfectly with the needs of PQ, due to constant amplitude frequency components associated with the main power signal or permanent distortions. These are easily identified and separated from frequency components associated with transient disturbances or with which amplitude fluctuates (so its source is not permanent). Using this method, PQ can be evaluated and used to find the detection of events or constant coupled frequency components.

In this method, PQ with SK methods have been developed [60, 86–89], one of them using implementation of the SK in an FPGA (Field Programming Gate Array) for a real-time signal analysis [56].

4.4 Harmonic Distortion

Harmonic distortion is a type of distortion caused when charge current does not follow voltage (a non-sinusoidal current). This effect was first observed in motors and transformers, due to nonlinearities of the magnetic circuit when it is over-fluxed or saturated. In those situations, current loses sinusoidal shape as the system saturates. However, the main source of harmonic distortion is nonlinear loads. In those kinds of loads, current is directly non-sinusoidal.

The presence of non-sinusoidal current creates a voltage deformation in generators, which is called voltage harmonic distortion. At the same time, when a system is fed with a non-sinusoidal waveform, it could create a higher deformation in the current, which implies a higher voltage harmonic distortion.

Harmonic distortion is the main frequency domain study in PQ in traditional indicators.

4.4.1 Types of Harmonic Distortion

Harmonic distortion is organised by frequency, with f_1 the power system frequency, as indicated in Table 4.1. DC is simply the constant value coupled to the signal.

4.4.1.1 Sub-harmonic Distortion

Harmonic distortion is the system reaction to a signal with the frequency f_1. Reactions have the same frequency or higher. However, it is possible to measure signals of frequencies less than the fundamental. As these signals have a source different from the fundamental, they are catalogued as sub-harmonics.

4.4.1.2 Harmonic Distortion

Traditional Fourier series theory indicated that any signal with a base frequency f_1 can be expressed with a set of terms with frequencies $f = hf_1$, where h is a positive integer. As the fundamental frequency, f_1, is the origin of the deformed signal, a deformed one has the same frequency, but with the harmonic distortion coupled.

Due to symmetry of the power system waveform, odd harmonics can be present (which change the shape but not the symmetry), but even

Table 4.1 Types of harmonics.

Type	Condition
DC	$f = 0$ Hz
Sub-harmonic	$f < f_1$
Harmonic	$f = h*f_1$ with h positive integer
Inter-harmonic	$f \neq h*f_1$ with h positive integer
Supra-harmonic	2 kHz $< f <$ 150 kHz

Source: Authors.

harmonics are very problematic (due to a change in the waveform shape and symmetry and a large effect on magnetic cores).

4.4.1.3 Inter-harmonic Distortion

With nonlinear loads, sometimes current shape cannot be perfectly decomposed into harmonic components ($f = hf_1$ with h a positive integer). Sometimes a discrete frequency or a frequency band appears in the frequency spectrum. These signals can be caused by a fundamental frequency, but with complex combinations of nonlinear systems.

4.4.1.4 Supra-harmonic Distortion

Signals can be found in power systems that are associated with power line communication (PLC), electromagnetic couples, frequency switching of converters, resonance of filters, etc. All of these kinds of signals are in a frequency range that traditionally has not usually been studied, but is now attracting more and more interest. All of these signals are becoming more and more common and should be studied in relation to their propagation, mitigation, interference and effects, as they are now being applied to all systems connected to the power network.

4.4.2 Sources of Harmonic Distortion

As indicated earlier, the first history of an harmonic source was magnetic cores (motors and transformers) when magnetic circuits were saturated. Then discharge lamps entered as a harmonic source; due to the arc re-start of each half-cycle, the returning zero current had a lower input amplitude. However, nowadays, power electronics introduce harmonics in a few different ways. First, rectifiers take current only in sinusoidal crests (when the input value is higher than the DC value), so current is similar to two pulses in a period, rather than a sinusoidal waveform. Another type of power electronic device that introduces harmonics is a motor starter, which only takes a percentage of the system waveform during a start.

Not only devices that take energy introduce harmonics and not all inverters deliver a perfect sinusoidal waveform, so some harmonics are introduced sometimes in the harmonic range, while others are in the supra-harmonic range. In addition, nowadays, there are more and more PLC systems, as domestic electric meters, which implies a supra-harmonic distortion.

Moreover, these are the known permanent harmonic sources, but any disturbance introduces effects over some frequencies, which can be measured as harmonic distortions.

4.4.3 Impact of Harmonic Distortion Over a Power System

Harmonic currents flow through conductors, but they cannot be used to deliver active power. They can be more or less present in the system, but at the moment they appear, they propagate using the network. These harmonic currents can be caused from:

- Increase of losses in magnetic cores (motors, transformers and generators) and reduction of its efficiency.
- Power factor reduction.
- Reduction of useful life of installation.
- Abnormal function or failure of equipment due to waveform distortion.
- Damage or destruction of components, as capacitors, e.g. for reactive power compensation.
- Coupled signals can cause interferences in the PLC or near communication lines.

The main consequence of harmonic currents is the loss of a sinusoidal waveform. This could sound trivial, but all definitions in relation to power systems are based on the assumption of a sinusoidal waveform (e.g. active and reactive power, power factor). As mentioned in previous chapters, researchers in the PQ field are demonstrating that these definitions need to be modified.

4.5 Traditional Theories of Electrical Frequency-Domain Indicators

4.5.1 Harmonic Measures

Regulations related to harmonic distortion measured in power systems are stated in IEC 61000-4-7 [84], and limits are set in IEC 61000-4-30 [1]. A brief explanation about harmonic measure follows.

As explained before, harmonic distortion is the presence of frequency multiples of fundamental one, so a procedure is set by regulation in order to measure properly frequencies multiples other than fundamental one, even when fundamental one is variable. Power system frequency is almost 50/60 Hz, but is never exactly 50/60 Hz and is always changing. Regulation 50160:2010/A1:2015 [13] sets a maximum variation of ± 0.1 Hz for interconnected systems and ± 2 Hz for isolated systems. This could seem to be low frequency variations, especially in interconnected systems, but up to 50 orders of harmonics are analysed, which implies a frequency change up to ± 5 Hz in interconnected systems and ± 100 Hz in isolated systems.

This implies that a double trouble, proper frequency must be measured in order to get the harmonic information, and DFT does not work well with frequencies that are not in bins. As explained before, DFT has a frequency resolution of $df = 1/T$, with T a segment length in seconds. Amplitude is taken for each frequency related with that resolution (DFT bins). However, what happens with frequencies in the middle of the bins? Ideally, there should be none, but when there are, the amplitudes of those frequencies are spread in surroundings analysed frequencies (bins), with the effect called spectral leakage. In practice, an experience involves introducing a sinusoidal signal of different amplitudes, with a sampling rate of 200 and taking 0.1 seconds.

This effect can be controlled with spectral windows where an amplitude modulation of time signals before applying DFT, but with consequences. Spectral windows limit spectral leakage to a few bins, but expand all frequency measures to a few bins. The number of bins depends on the type of spectral window. In Figure 4.1 can be seen an example of spectral leakage and its correction using a Hanning window. A 50 Hz centre example has been given in order to compare the effects, but frequency changes over higher-order harmonics have been represented with frequency changes of up to 5 Hz.

It can be seen that for power system frequency in interconnected systems, the effect is not very important. However, for isolated systems, or for higher-order harmonics, the effect can be seen over a few surrounding frequencies. However, with the Hanning window, the response, even for 51 Hz, is very similar, with values in three bins. The spectral resolution is 10 Hz, so the worst situation is 55 Hz, a signal with a frequency in the middle of two bins.

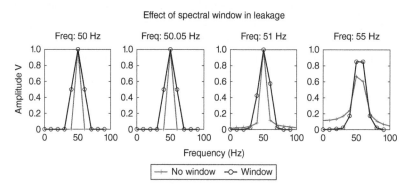

Figure 4.1 Effect of intermediate frequencies with and without spectral windows. *Source:* Authors.

Even in that situation, windowing recovers part of the signal amplitude and limits leakage to 4 bins.

Spectral leakage only appears when there are signals out of bins. In order to avoid this, the regulations stipulate a procedure that adjusts the sampling rate in order to show that the power system frequency exactly matches the bin frequency. As harmonics components have frequencies multiple of the fundamental one, that will also match the bins, thus avoiding spectral leakage in harmonics. If the system is not synchronised (sampling frequency is not adjusted), the Hanning window must be applied and data must be marked as not synchronised.

That turns the DFT axis from a frequency axis to a harmonic order axis, where the exact frequency of each bin changes continuously. According to the regulations, for spectral measures it must be taken as 10/12 cycles for 50/60 Hz signals, which implies a duration of 0.2 seconds, so that $df = 1/T = 1/0.2 = 5\,\text{Hz}$. Additionally, a calculation is made for each harmonic amplitude, with a quadratic sum of bins, centred in harmonics, making harmonic groups. For inter-harmonics, groups are taken among the harmonic bins. As a result, harmonic and inter-harmonic readings are taken.

4.5.2 DFT-Derived Measures

Once the frequency composition is decided, there are some calculations that can be done in order to obtain a simple metrics of the harmonic composition. The first one is to indicate harmonic amplitudes as a relative amplitude of the fundamental one, thus providing a metrics of how they are affecting the waveform individually. Regulation IEC 61000-4-30 [1] establishes a limit for each individual harmonic and for its combination. In the harmonic measure, this is done in relation to harmonic groups. Some bins around the frequency are taken according to calculations indicated in IEC 61000-4-7 [84]. Hereafter, harmonic groups will be called harmonics.

Total harmonic distortion (THD) is a measure of all harmonics present in the waveform, calculation as shown in the following equation:

$$\%\text{THD} = \frac{\sqrt{\sum_{h=2}^{\infty}\left(M_h\right)^2}}{M_1}\cdot 100 \tag{4.1}$$

where M_h is the RMS voltage or current for the h harmonic and $h = 1$ is the fundamental harmonic associated with the fundamental frequency. Usually, this calculation is taken up to the harmonic order 25, but some regulations stipulate taking it up to harmonic order 50. Sometimes, when the THD of the voltage waveform is calculated, this is indicated by THDV and where THD of the current is calculated, it is indicated as THDI.

The same definition can be used for inter-harmonics, taking the total inter-harmonic distortion (TIHD), where inter-harmonic groups are used instead of harmonics groups, after taking a measure of how distorted the waveform is. The calculation is shown by

$$\%\text{TIHD} = \frac{\sqrt{\sum_{h=1}^{\infty} \left(M_{i,h} \right)^2}}{M_1} \cdot 100 \tag{4.2}$$

Now, inter-harmonics groups, starting from the first one are used for calculated the quadratic sum, and fundamental harmonic are used as denominator.

Another type of distortion indicated before is sub-harmonic distortion, and that is indicated with the total sub-harmonics distortion (TSHD), as indicated by

$$\%\text{TSHD} = \frac{\sqrt{\sum_{h=1}^{s} \left(M_{s,h} \right)^2}}{M_1} \cdot 100 \tag{4.3}$$

4.6 HOS Contribution in PQ in the Frequency Domain

DFT and derived measures are an indication of waveform distortion, where the RMS of each harmonic is measured. However, it is an averaged information of a snapshot. If information about the evolution of harmonics need to be obtained, 2D graphs (spectrogram: all harmonics vs. time) or other representations are required where amplitude or power evolution are observed.

In the time domain, HOS has been really useful in identifying frequency and waveform changes, but in the spectral domain, frequency is fixed to the associated bin and the shape is always sinusoidal (by definition of a spectral component in Fourier series theory). HOS in the spectral domain has a different use. SK uses some DFT for calculations and returns kurtosis of a point distribution of each spectral component. They are sinusoidal, but as some segments are taken together, completely analysed signals are perfectly sinusoidal if the amplitude remains constant. For any amplitude change, variations in point distributions can be observed.

Finally, in the fourth-order spectral domain, throw SK is an indicator of the amplitude evolution of each spectral component, and can either be a harmonic frequency or not, which indicates −1 for a constant amplitude.

Even when only noise behaviour is measured, an SK value around 0 is observed. On the other hand, when fast and sudden changes occur, a high SK value can be observed.

4.6.1 Spectral Kurtosis

As indicated before, SK is an indirect calculation of the fourth-order spectral domain, using M non-overlapped segments or realisations of the signal under interest, and calculating DFT over them. As SK is a calculation over M frequency-domain data, that data must be associated with the same frequency axis, keeping the sampling frequency and segment length constant and refusing the sync procedure stipulated for harmonic measures. This implies application of the Hanning window in order to avoid or limit the spectral leakage. Taking M non-overlapped realisations implies a need to use $T = 0.2$ second as the signal length. Once the signal is segmented in realisations and DFT applies to them, SK is calculated using the following equation:

$$\hat{k}_x\left(m\right) = \frac{M}{M-1}\left[\frac{\left(M+1\right)\sum_{i=1}^{M}\left|X_i\left(m\right)\right|^4}{\left(\sum_{i=1}^{M}\left|X_i\left(m\right)\right|^2\right)^2} - 2\right] \tag{4.4}$$

where $X_i(m)$ is the amplitude for the m bin of the DFT of the signal segment i. To calculate SK $\hat{k}_x(m)$ of each bin (each frequency), the amplitude given for DFT of all realisations for that specific bin (frequency) is used. $M/(M-1)$ is a bias correction for calculation with a low number of realisations.

DFT is usually calculated using FFT, due to a reduction in computational cost. However, FFT needs a power of 2 in the number of points. Some systems require this specific point length (mainly low-level processors); others work faster with that number of points.

In order to understand the SK values better, kurtosis calculation (in the time domain) is now going to be explained, together with its application over sinusoidal waveforms. This is an example of kurtosis of a single spectral component, which is each value of SK.

4.6.1.1 Kurtosis

Chapters 2, 3, and 5 introduce methods that implement normal kurtosis in the time domain. Here, kurtosis is introduced in order to understand the kurtosis concept applied to frequency domain analysis.

In probability theory and statistics, kurtosis is a measure of the 'tailedness' of the probability distribution of a real-valued random variable. Tailedness is a measure of the relative quantity of points in distribution tails. This definition needs another explanation.

A time series is analysed, which is a sequence of values with a time mark. However, if all the values are taken together, as a random variable, concepts related to probability theory and statistics can be applied. This is the point of view of a statistical analysis of a waveform, where data is considered as a random variable.

The analysis focuses on the probability density function (PDF) and its properties. The PDF indicates the probability of taking each value and then taking a random number from the random variable. Then a random distribution is considered (normal, Gaussian, Weibull, T-Student, etc.). The PDF is defined by a parametric equation. However, when experimental data are considered, only an experimental PDF (ePDF) can be obtained.

Kurtosis, as other statistical moments, is a description of PDF or ePDF and is specifically a description of the number of points in the tails. The proper definition of a tail in a PDF is an unbounded limit for positive or negative values (a positive or negative tail), with a really low possibility. When the PDF is bounded (positive, negative or both), there are no tails. This situation does not render kurtosis useless; it just needs a clarification. Kurtosis is a measure of points relative to position around the mean value, giving an indication of a high concentration around the mean value, values spread in a range, values concentrated far from the mean value, etc. Figure 4.2 shows a first example of the kurtosis value.

Figure 4.2 shows some statistical distributions, with the same mean value and different point spreads. It is easy to see that the uniform distribution (U), with the same probability for all values, shows the lowest value in this representation, and the Laplace distribution (D) shows the highest kurtosis value, with more probability to take a point just in the mean value than in any other value.

It is not a good practice to calculate ePDF for kurtosis, due to some approximations that should be done and because there are direct ways to calculate kurtosis of time series. Kurtosis is the fourth-order centred moment, with zero lag, as seen in the following equation:

$$\text{kurt}\left[X\right] = \frac{E\left[\left(X-\mu\right)^{4}\right]}{\left[E\left[\left(X-\mu\right)^{2}\right]\right]^{2}} - 3 \tag{4.5}$$

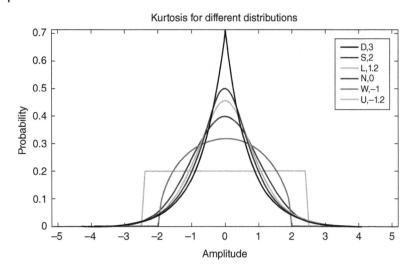

Figure 4.2 Comparison of kurtosis of different distributions. *Source:* Authors.

where X is the data under test, $E[\ ]$ is the expected value and μ is the mean value of X. As seen in Figure 4.2, this calculation returns a kurtosis value of 0 for normal or Gaussian distributions (N), higher values for distributions with points more concentrated around the mean value and lower kurtosis values for distributions with points less concentrated around the mean value. The proper name for this calculation is excess kurtosis (hereinafter kurtosis), which is the traditional kurtosis −3. Therefore, it is possible to find kurtosis 3 for Gaussian distribution, but it is not the usual representation. SK values are related to excess kurtosis.

Before making a start with sinusoidal signals, two extreme situations will help in an understanding of extreme kurtosis values. First, we are going to focus on the lowest kurtosis value. As seen before, kurtosis shows a lower value where points are farthest from the mean value. Therefore, if all points are concentrated in both tails, at the same distance over and under the mean value, the mean values will be just in the middle and at a distance where all points are at a maximum. In any other situation, the mean value moves and the distance is lower to one point group. This can be obtained with a square waveform; Figure 4.3 shows an example.

A waveform and histogram can be seen, and it is easy to see that the mean value is 0 and all points are at one value distance (±1). However, that distance is considered as squared or to the fourth power, so $(X-\mu)^4$ or $(X-\mu)^2$ are used for all series. Therefore, the expected values $E[(X-\mu)^4]$ and

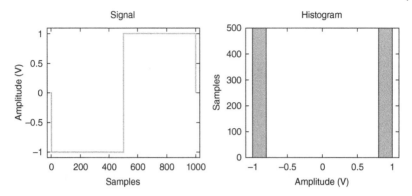

Figure 4.3 Example of a square signal, the lowest kurtosis situation, −2 value.
Source: Authors.

$[E(X-\mu)^2]^2$ are for all one data vector. Calculations are done in the following equation:

$$\text{kurt}_{\min}[X] = \frac{E\left[(X-\mu)^4\right]}{\left[E\left[(X-\mu)^2\right]\right]^2} - 3 = \frac{1}{1} - 3 = 1 - 3 = -2 \qquad (4.6)$$

Therefore, it is shown that the minimum value that the kurtosis can be taken is −2.

We will now focus on the opposite situation, the maximum kurtosis value. As seen previously, the high concentration of points around the mean value implies high kurtosis values. In order to obtain an extreme situation, a signal is generated with all points except one having the same value. As all points have the same value, the range of distribution is limited to that value, and a minimal noise (even computational noise) creates a uniform distribution. However, with an outlier, the range expands, making calculations safer. Figure 4.4 shows an example of this experience.

A thousand values of vectors have been considered, where all have a value of 0 except one, with a 1 value. First, the mean value of these data is needed, which is calculated below:

$$\bar{X} = \frac{1}{1000} = 0.001 \qquad (4.7)$$

A single one value for over a thousand values does not change the mean value significantly and would be considered as zero, in order to make a fast estimation of the kurtosis value. If the mean value is zero, $(X-\mu)$ is zero,

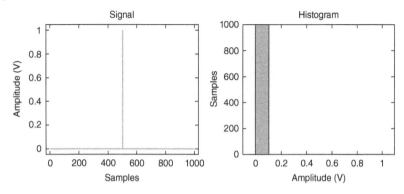

Figure 4.4 Example of an impulsive situation, the highest kurtosis situation, 997 value. *Source:* Authors.

except for the outlier value. Therefore, the kurtosis calculation only considers one point of 1000, remembering that $E[X] = \sum X/N$. The following equation shows the calculations:

$$\text{kurt}_{max}\left[X\right]=\frac{E\left[\left(X-\mu\right)^{4}\right]}{\left[E\left(X-\mu\right)^{2}\right]^{2}}-3=\frac{E\left[\left[1\right]^{4}\right]}{\left[E\left[1\right]^{2}\right]^{2}}-3=\frac{\dfrac{\left[1\right]^{4}}{1000}}{\dfrac{\left[1\right]^{4}}{1000^{2}}}-3=1000-3=997$$

(4.8)

It can be seen, and considering that N (number of points) is 1000, that the maxima kurtosis value is $N-3$ for impulsive behaviours (for an outliers presence with a relative amplitude higher than other values). The proper calculation of kurtosis over the data, considering the variation of the mean value, returns a kurtosis value of 995. With that value, our approximation has been acceptable.

With all this, a question is mandatory; if an outlier affects the outcome in this way, would more than one affect in the same way or differently? Figure 4.5 shows an experience with up to half a signal with the amplitude changed.

If more than half of the waveform presents a change in amplitude towards the same value, the changed amplitude actually turns the new normal amplitude and the old normal amplitude into the outlier amplitude. Indeed, for 500 changes, the amplitude takes half the time for each value and has a square waveform. Kurtosis returns in this situation -2, the one associated with a square waveform. With k (points affected by amplitude change) of 277, the -1 kurtosis value is observed and kurtosis shows a 0 value for

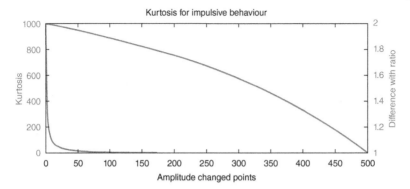

Figure 4.5 Example of an impulsive situation, for different numbers of amplitude changed points. *Source:* Authors.

$k = 173$. However, even starting at 996, with only $k = 10$, kurtosis takes the value 95.

All this proves the fast reduction as increases in the number of outliers, and there is a good relation between our estimation and the real kurtosis for impulsive behaviour. In all data, time series kurtosis shows a value underestimated by $M/k - 3$, but is really consistent and is between 2 and 1 (due to the change of the mean value). Therefore, for low numbers of outliers, there are high kurtosis values, the error between 1 and 2 is not significant and $M/k - 3$ is a good estimator.

4.6.1.2 Single Spectral Component Kurtosis

Once the kurtosis concept has been explained, it will be the focus for sinusoidal signals, in order to understand SK values. First, the constant amplitude sinusoidal signal will be studied in Figure 4.6.

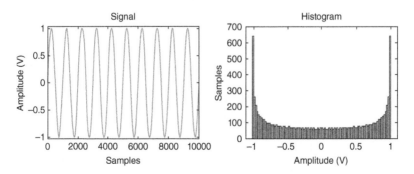

Figure 4.6 Constant amplitude sinusoidal signal and histogram. *Source:* Authors.

In Figure 4.6, the waveform and data histogram are shown. The point distribution is bi-modal (the two most common values) in extremes values (±1), with a higher point concentration as the amplitude is near to the extreme amplitude and there is less point concentration around zero (the mean value). This point distribution is a mix between the square signal and uniform distribution, so a kurtosis for this signal is −1.5, between the ones associated with uniform distribution and the square signal.

Previously, the SK value for a constant amplitude spectral component has been indicated as −1, but the kurtosis value for a sinusoidal waveform is −1.5. Studies covering sinusoidal waveforms will be done in the future and this discrepancy will be studied in detail.

This is the base response for a constant amplitude sinusoidal signal, which is a constant amplitude spectral component. Now, an SK will be calculated with $M = 10$ (in this experience 10 cycles), with all cycles having the same amplitude except one, which shows an amplitude 10 000 times higher. First, the signal and histogram are shown in Figure 4.7.

The amplitude of the last cycle is too high and other cycles cannot be observed in the waveform. The histogram reveals the changes in the point distribution. As one cycle with a really high amplitude is present, points associated with the other cycles can be observed to be nearer to the mean value. In other words, a higher point concentration around the mean value can be observed. This changes the kurtosis value completely, obtaining 12.015. This value would not have any sense at this point, but another experience is performed, taking different numbers of realisations (cycles) and a single realisation with the amplitude changed. Table 4.2 shows the simulation results.

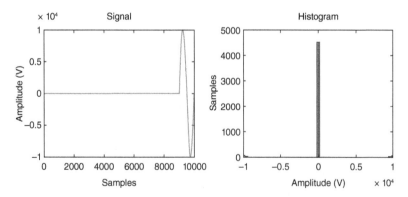

Figure 4.7 Impulsive amplitude change sinusoidal signal and histogram. *Source:* Authors.

Table 4.2 Different situations of impulsive behaviour.

Cycles (M)	10	25	50	100	250	500	1000
Excess kurtosis value	12	34.5	72	147	372	747	1497
Normal kurtosis value	15	37.5	75	150	375	750	1500
Ratio: kurtosis/M	1.2	1.38	1.44	1.46	1.49	1.49	1.49
Ratio: normal kurtosis/M	1.5	1.5	1.5	1.5	1.5	1.5	1.5

Source: Authors.

Table 4.2 shows excess kurtosis values obtained for all situations. In addition, the third row represents normal kurtosis (an excess one +3). This is done to obtain an easy understatement of the results. When the ratio between the normal kurtosis and the number of realisations and the excess kurtosis and the number of realisations is found, it has been observed that the ratio between normal kurtosis and the number of realisations is constant for all situations. This relation is normal kurtosis/$M = 1.5$ or, in other words, normal kurtosis $= 1.5M$. However, the SK returns excess kurtosis and its ratio is not consistent. For that reason, a ratio with normal kurtosis can be done, with a correction done at the end. Thus, excess kurtosis (kurtosis) $=$ normal kurtosis $- 3 = 1.5M - 3$. These considerations can only be applied under impulsive behaviour; it is a huge increase of amplitude of some cycles.

Using this, the effect of a different number of cycles with the amplitude changed is studied in the following experience, taking 1000 realisations of unitary amplitude sinusoidal signals, one cycle for each one, and considering from 1 to 7 changed amplitudes of sinusoidal cycles up to 10 000. Table 4.3 shows the simulation results.

This experience only considers kurtosis (excess kurtosis), as this is the one related to SK. Now the kurtosis value returned at analysis of the data vector is compared with the kurtosis estimator stipulated in the previous experience. Table 4.3 shows that the estimated kurtosis value divided by the

Table 4.3 Different number of impulsive sinusoidal cycles situation.

Cycles with higher amplitude (k)	1	2	3	4	5	6	7
Kurtosis value	1497	747	497	372	297	247	211.2
Ratio: 1.5 * M/kurtosis	1	2	3	4	5	6	7

Source: Authors.

measured kurtosis value returns the number of affected cycles. This implies that the maxima kurtosis returned is reduced by a factor of k, where k is the number of cycles with the amplitude changed.

Reformulating the ratio shown in Table 4.3 gives

$$1.5M / \text{kurtosis} = k \tag{4.9}$$

and a new kurtosis estimator can be obtained, including a number of affected cycles:

$$\text{kurtosis} = 1.5M / k \tag{4.10}$$

Up to this point, a huge change of amplitude has been performed. With lower amplitude changes, kurtosis is able to make softer changes. In order to evaluate the effect of the amplitude of the changed cycle in a kurtosis value, a simulation is done using unitary sinusoidal cycles, and then the amplitude of the changed cycles have different amplitudes, from 1 to 500. A combination of amplitude changes with the number of realisations is done, changing one cycle under a different number of realisations (cycles) and combining them with a number of affected cycles produces 1500 realisations from this experience. Figure 4.8 shows the results associated with those experiences.

The kurtosis value is conditioned by the number of realisations or the number of cycles affected by amplitude change. This is shown in Figure 4.8, where the final value change is higher as more realisations are taken and lower as more cycles are affected. However, the final kurtosis value seen in previous experiences for high amplitude changes need certain amplitude changes. This can be observed in graphs where the amplitude change

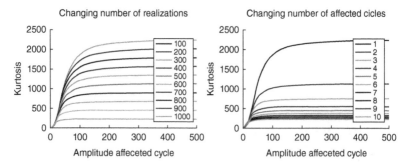

Figure 4.8 Kurtosis for different amplitudes of an affected cycle, in one affected cycle, changing total cycles (left) and 1500 total cycles and changing the number of affected cycles (right). *Source:* Authors.

needed to reach the final kurtosis value is similar, no matter whether the maximum kurtosis value is limited by the number of realisations or the number of affected cycles.

Even having fourth-order outliers must have enough weight to return a high kurtosis value. As more points are around the mean value (this implies less points as outliers and less numbers of affected cycles), affected cycles must have higher amplitudes. In this way, the relative weight of a few points that are further from the mean value are enough to compensate for the relative weight of many points near to the mean value. As more points are considered as the amplitude changes, more relative weights are given to them, even with less amplitude, and the maximum kurtosis value can be reached in this way with lower amplitude changes.

Extreme situations (constant and impulsive changes) have been studied. However, in between these, there are many different behaviours. As an example, a soft amplitude variation is shown in Figure 4.9, which considers 1000 cycles, from amplitude one to two.

The histogram now shows a softer situation. Extreme values do not have a maximum number of points and this maxima concentration is ± 1 (the initial amplitude). A softer distribution in point density can be observed, joined to a 'round' shape of the histogram. It is an increase from a minimum extreme value to a maximum of ± 1. This point distribution returns a kurtosis value of -1.2918, near to the uniform distribution. In this middle situation, kurtosis of a spectral component is indicative of how far a signal is from a constant amplitude.

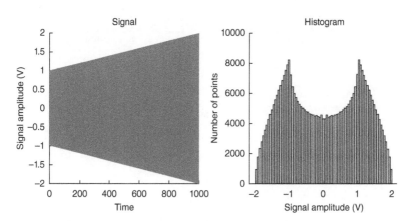

Figure 4.9 Sinusoidal signal with a linear amplitude double (left) and a histogram of this signal (right). *Source:* Authors.

4.6.2 Spectral Kurtosis Basic Usage

Once SK values have been explained, an SK of a 50 Hz sinusoidal signal (a perfect European power signal), with 50 dB SNR (signal to noise ratio) of coupled Gaussian noise, taking 500 realisations and with an Fs of 1024 samples/s, is analysed. Figure 4.10 shows its SK.

Figure 4.10 shows the response studied for 50 Hz, −1 kurtosis value and random values, around 0, for all other frequencies. Real-life signals always have any level of noise coupled, which has been represented with an SNR of 50 dB. This level has been selected for an SNR when a power network from 50 to 70 is measured. Even taking the highest noise level considered for the power system, the response for a constant amplitude is as explained previously.

In this experience, two concepts need a deep study: one kurtosis value for a constant amplitude and oscillations around zero in only the noise frequency. These two behaviours are going to be studied as changing numbers of realisations and noise. Figure 4.11 shows the results of these experiences.

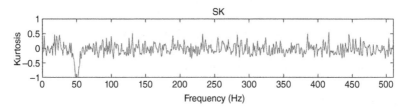

Figure 4.10 SK of 50 Hz sinusoidal waveform, with constant amplitude and noise coupled. *Source:* Authors.

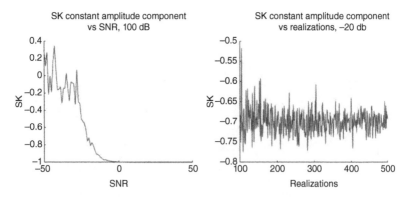

Figure 4.11 SK response for constant amplitude spectral component, with different noise levels (100 realisations) and different number of realisations (SNR of −20 dB). *Source:* Authors.

First, the SK responses for a constant amplitude spectral component with different noise levels are tested. Graphs show a perfect −1 value for SNR 0 or higher and perfect values for SNR 50 dB as related to the power network. For examining the effect of a number of realisations, an SNR in which the response is partially affected is selected as −20 dB, where the response is around −0.8. With this noise level, the number of realisations are changed from 100 to 500. A value of around −0.7 can be seen in any number of realisations.

With that, the SK response for a constant amplitude spectral component is affected by the noise level, but it is not affected by the number of realisations.

Then an experience relating to the SK values observed for noise only was carried out using spectral components. During this experience, maximum and minimum values for oscillations were observed in noise only spectral components being saved for different noise levels and, as done before, with −20 dB SNR, with different numbers of realisations. The results are shown in Figure 4.12.

This experience shows a behaviour completely opposite to the previous one. The oscillation level is not related to the noise level and changes with the number of realisations, showing lower oscillations as the number of realisations increases. Indeed, there is a relation between those oscillations and the number of realisations, as postulated in reference [102], indicated by

$$\text{var}\left\{\hat{k}(m)\right\} = \frac{4M^2}{(M-1)(M+2)(M+3)} < \frac{4}{M} \qquad (4.11)$$

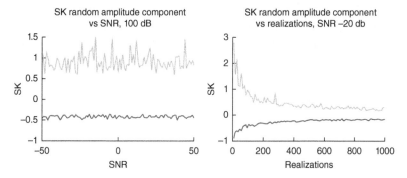

Figure 4.12 SK base oscillations (maximum and minimum values), with different noise levels (100 realisations) and a different number of realisations (SNR of −20 dB). *Source:* Authors.

This expression indicates the SK variance for noise only spectral components and it can be seen that, in the complete expression and in a simplified expression, it is only related to *M* (the number of realisations).

The next experience studies the opposite behaviour, impulsive amplitude change. Now the same signal is used as in the previous experience, with a perfect sinusoidal waveform of 50 Hz and a Gaussian additive noise with an SNR of 50 dB, but now, in one realisation, another sinusoidal signal with a unitary amplitude is included – in this situation with 150 Hz. The SK of that signal is shown in Figure 4.13.

Figure 4.13 shows a perfect impulsive response for the frequency under test, but the value relating to the impulsive behaviour is 100. In a kurtosis analysis of a sinusoidal signal, a value of around 147 (1.5*M*/k − 3) has been obtained. The same value escalation can be seen in the constant amplitude frequency component, where −1 SK values are obtained and a −1.5 kurtosis value is associated with the sinusoidal signal. This is due to the SK value and is related to kurtosis, but when a scale factor of 1.5 is applied, the value is reduced and a minimum of −1 and a maximum of the number of realisations is obtained.

In a kurtosis study of sinusoidal signals, a huge amplitude was needed for a high kurtosis response, but now a unitary amplitude sinusoidal signal created an impulsive behaviour. This is due, before introducing the unitary signal, in a specific spectral component, which is in the noise level and the increased ratio in relation to the noise was higher, as seen in sinusoidal signals. Indeed, the SNR of 50 dB is a measure relative to the main signal (50 Hz constant amplitude). First, the SNR is studied in order to determinate the noise level:

$$SNR_{dB} = 10 \log_{10} \frac{\sigma_{signal}^2}{\sigma_{noise}^2} \tag{4.12}$$

As a spectral component is a perfect sinusoid, its variance is known and has a relationship with its amplitude, $\sigma_{signal}^2 = 0.5A^2$. Using this, the previous equation can be expressed as

$$\frac{SNR_{dB}}{10} = \log_{10} \frac{0.5A^2}{\sigma_{noise}^2} \tag{4.13}$$

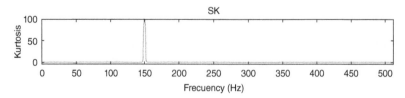

Figure 4.13 SK for an impulsive situation. *Source:* Authors.

Removing the logarithm powered by 10 and reorganising, the noise variance can be obtained as seen in the following equation:

$$\sigma^2_{noise} = \frac{0.5A^2}{10^{\frac{SNR_{dB}}{10}}} \tag{4.14}$$

For a unitary amplitude with an SNR of 50 dB, the noise variance is $5 \times 10^{-6} V^2$. For any amplitude of an impulsive spectral component, the same noise is present and causes different noise conditions in relation to the signal. The same situations occur for constant amplitude detection, but in that situation the SNR from 0 dB is enough to evaluate a constant amplitude, as seen previously. In relation to the change of SK in impulsive behaviour spectral components, finding the different noise levels and experience is calculated by using the SNR in relation to the impulsive signal amplitude. The SK values and SNR are shown in Figure 4.14.

For an SNR higher than 20, the response is 99.5% of the final one. In this situation, 1000 realisations occur where the SK final value is higher than the previous 100 realisation simulations, but the evolution is the same. In the following equation the amplitude needed in a specific spectral component to reach a specific SNR, in a signal with a SNR in relation with the main signal, is studied using the following equations:

$$\sigma^2_{noise} = \frac{0.5A^2_{Main}}{10^{\left(\frac{SNR_{dB\_Main}}{10}\right)}} \text{ and } \sigma^2_{noise} = \frac{0.5A^2_{Comp}}{10^{\left(\frac{SNR_{dB\_Comp}}{10}\right)}} \tag{4.15}$$

$$\text{where } \frac{0.5A^2_{Main}}{10^{\left(\frac{SNR_{dB\_Main}}{10}\right)}} = \frac{0.5A^2_{Comp}}{10^{\left(\frac{SNR_{dB\_Comp}}{10}\right)}} \tag{4.16}$$

and

$$A^2_{Comp} = \frac{10^{\left(\frac{SNR_{dB\_Comp}}{10}\right)}}{10^{\left(\frac{SNR_{dB\_Main}}{10}\right)}} A^2_{Main} = 10^{\left(\frac{SNR_{dB\_Comp} - SNR_{dB\_Main}}{10}\right)} A^2_{Main} \tag{4.17}$$

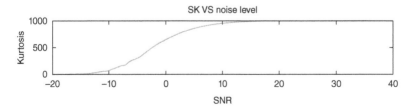

Figure 4.14 SK max for an impulsive situation, changing SNR. *Source:* Authors.

$$A_{Comp} = \sqrt{10^{\left(\frac{SNR_{dB\_Comp} - SNR_{dB\_Main}}{10}\right)} A_{Main}^2} \tag{4.18}$$

From these equations, it should be possible to calculate the minimum amplitude of the spectral components. The constant amplitude and impulsive amplitude are calculated for unitary amplitude signals and 50 dB SNR as follows:

$$A_{Constant} = \sqrt{10^{\left(\frac{0-50}{10}\right)} \times 1^2} = \sqrt{10^{-5} \times 1} = 0.0032 \tag{4.19}$$

$$A_{imp} = \sqrt{10^{\left(\frac{20-50}{10}\right)} \times 1^2} = \sqrt{10^{-3} \times 1} = 0.0316 \tag{4.20}$$

With lower noise contamination (a higher noise in the range considered for a power system has been used), lower values can be obtained. This implies a single signal in one realisation with an amplitude of 3% in relation to the fundamental one, which would be detected as impulsive (averaged amplitude in realisation), and spectral components with constant amplitudes from 0.3% in relation to fundamentals would be detected properly.

A last consideration must be studied. Up to this point, amplitude values per frequency have been considered. This implies a constant amplitude inside each realisation, because in the DFT model, the time domain signal in the frequency domain and amplitude changes in the analysed segment involve more than one frequency, which is frequency leakage. Therefore, if an amplitude change occurs inside a realisation, a leakage can be observed, involving more frequencies. SK considers values for those other frequencies, in relation to additional data from other realisations. In order to illustrate this, the same impulsive example as previously is used, with an analysis of 100 realisations where one realisation is affected, but now the signal is only present in half of the realisation with an amplitude of 0.05. Figure 4.15 shows the SK for this signal.

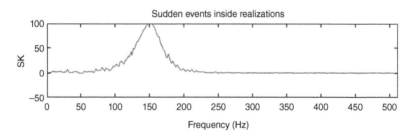

Figure 4.15 SK for a sinusoidal signal during part of a realisation. *Source:* Authors.

Leakage can be observed around the signal introduced. Even with leakage control of the Hanning window, leakage in one realisation produces an impulsive behaviour. This effect is really important in a real signal analysis, due to fast level changes or temporary signals, due to the fact that they occur at any moment, and not at the beginning or end of a realisation.

4.6.3 Spectral Kurtosis and Power Quality

Once SK usage has been explained and its values have been understood, an application concerning the power system can be explained. First, an aspect related to the state of the power system is going to be studied. As previously indicated for PQ, the SK are analysed in realisations of 200 ms length, according to regulations, and with application of the Hanning window due to constant frequency, sampling is used. This implies a frequency resolution of 5 Hz.

SK can study constant amplitude spectral components, which can be applied to two aspects, stability of the fundamental frequency component and detection of permanent harmonic distortion.

In voltage waveforms, amplitude of the fundamental frequency component is usually almost constant, but sometimes it fluctuates. This voltage-level stability can be evaluated easily using SK, due to a perfect constant amplitude that implies a −1 SK value, and fluctuations increase this value.

Harmonic distortion can be classified as permanent harmonic distortion (does not change or changes slowly cycle by cycle) and transient harmonic distortion (change cycle by cycle). The first one is detected as a constant amplitude frequency component, if it has an amplitude higher than 0.3% more than fundamental. Returned values would be −1 or a bit higher according to the strictly constant amplitude or a bit of change. The transient harmonic distortion appears after an operation and is the result of connect or disconnect systems (filters, non-linear loads, capacitors), even after some disturbances. After few seconds it mitigates. Depending on their duration, they are being seen as impulsive or low variation. Figure 4.16 shows some of these situations.

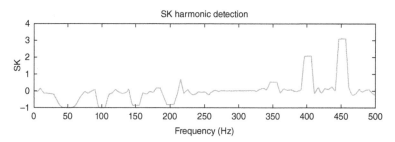

Figure 4.16 SK for different harmonic distortions in a power waveform. *Source:* Authors.

The 50 Hz harmonic shows the perfect −1 for an always constant spectral component. The second harmonic (100 Hz) has a constant amplitude, with an averaged amplitude of 0.05, shows −1, and so the constant amplitude behaviour has been detected perfectly. The third harmonic (150 Hz) can double its amplitude, with an average amplitude of 0.075. This gives an SK of −0.85, as expected for a spectral component with a double-amplitude (the $−1.29/1.5 = −0.86$ seen in the sinusoidal signal situation with the 1.5 correction returns $−1.29/1.5 = −0.86$). The fourth and fifth harmonics have the same relative amplitude change but a different averaged amplitude, being 0.00075 for the fourth harmonic and 0.000075 for the fifth one. The fifth one is under detection for a constant amplitude.

Harmonics from the sixth to the ninth have a similar behaviour. They are sinusoidal signals with 0.05 amplitude, which are only present during part of the signal time. The sixth harmonic is present during half of the simulation time, and the duration of each harmonic decreases, ending with one-fifth of the simulation time in the ninth harmonic. Impulsive behaviour is only a small piece of the signal with higher amplitudes. As the initial amplitude is noise, 0.05 is really a higher amplitude and a response would be obtained for any higher amplitude. It has been seen that if lower time harmonic is present, a higher SK value is obtained.

This detection is even done when a leakage is produced, when harmonics start or end in the middle of a realisation. The same simulation is done with a start delay of half a realisation, and the result is shown in Figure 4.17.

Leakage is produced around introduced frequencies, but the values associated with introduced frequencies are not affected. Table 4.4 shows SK values for both simulations, with and without leakage, with only one change, because in leakage harmonics start and end in the middle of a realisation, and takes one realisation less (half at the start and half at the end). Therefore, non-leakage simulation can be done by removing one realisation of the

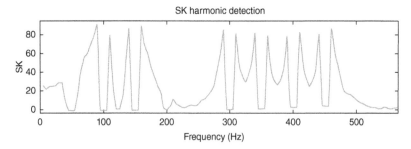

Figure 4.17 SK for different harmonic distortions in a power waveform, under leakage conditions. *Source:* Authors.

Table 4.4 Different number of impulsive sinusoidal cycles situation.

Frequency	SK normal	SK leakage
100	−0.9434	−0.9452
150	−0.7987	−0.7988
200	−0.1398	−0.2809
250	−0.0402	7.2798
300	0.2469	0.2534
350	0.8902	0.9041
400	3.0410	3.1061
450	4.7016	4.8294

Source: Authors.

harmonics in any situation. For example, constancy is now 99/100 realisations, but due to leakage it is $98 + 1/2 + 1/2$ (the start and end). Therefore, SK can measure the same harmonic presence, but in a different position.

Nevertheless, 200 and 250 Hz are the lowest amplitude values, while all other spectral components return similar SK values, depending on leakage behaviour. This shows that, even with leakage, SK returns the value associated with harmonic behaviour on each spectral component, and leakage does not affect the SK measurement. Moreover, it gives additional information about the relative position of disturbance in relation to realisations. It is important to understand that important data is the frequency value SK, not the leakage SK value.

Now two power system disturbances will be studied. The first one to consider is the most similar to harmonic temporal distortion, oscillatory transiency. This disturbance consists in high-frequency oscillations around the power waveform, with a fast amplitude start and exponential decay. Figure 4.18 shows the result of an example of analysis of this kind of disturbance.

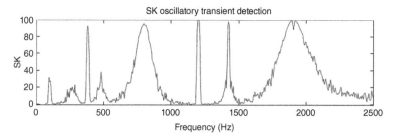

Figure 4.18 SK for different oscillatory transient conditions. *Source:* Authors.

In this simulation, different conditions have been introduced with each frequency, such as changing, start point, duration and frequency. Frequency has no effect over SK, due to DFT detection in the same way amplitude has for any spectral component. A maxima amplitude of the transient has been set at 0.1 for all transients. Exponential decay is calculated to achieve 5τ at the final time, when oscillations end.

The first oscillatory transients, with an oscillation frequency of 70 Hz and a duration of 1 cycle, are all inside a realisation. It cannot be seen because the spectral measure is too small, the exponential decay reduces the amplitude too quickly and an average of over 10 cycles (200 ms realisation) lowers the noise level amplitude. Then two three cycle length transients are introduced, the first with 100 Hz, all inside a realisation, and the second one with 270 Hz, split into two realisations by its first cycle. For 100 Hz, a response similar to impulsive harmonics, but wider, can be seen. An amplitude constant change inside realisation creates this leakage around the frequency of interest. On the other hand, the SK for 270 Hz has a lower value and a higher leakage. As disturbance covers two realisations, a less impulsive behaviour is detected. In addition, the split signal involved more frequencies, creating a wider spectral leakage.

Then two groups are considered (the first group with lengths of 5 cycles and the second group with lengths of 10 cycles). In both groups, the first signal is all inside one realisation, the second one has been split by its first cycle, and the third one has been split by its second cycle. It can be seen that a consistent response for oscillatory transients is all in one cycle, with an SK peak and a small amount of leakage, higher than seen in temporary transients. The more split among realisations the transient is, the more leakages can be observed, reaching wider domes, as seen in temporal harmonics. Lastly, leakage can be observed in more frequencies in SK than in harmonics, showing an increase of SK for many frequencies. So even when there is a reduction of SK in the dome, wider domes (zones of increased SK) are related to oscillatory transients.

Previously, it has been postulated that DFT could read amplitude perfectly for any frequency, but only multiple frequencies rather than frequency resolution have been used, in order to examine other characteristics perfectly. Now frequencies that are not multiples of frequency resolution are going to be used, with a 3-cycle length, a transient in one realisation and 0.1 of maxima amplitude. Figure 4.19 shows the result of different oscillation frequencies, with a difference of 100.5. As the frequency resolution is 5 Hz, this makes the most of the frequencies that are not coincident with DFT bins.

Figure 4.19 also shows a similar response, whose differences are associated with random noise, so oscillation frequency does not affect the SK measure,

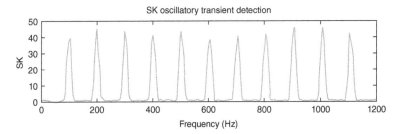

Figure 4.19 SK for different oscillatory transient conditions, with frequencies multiples of the frequency resolution. *Source:* Authors.

which was a frequency coincident or not with a DFT bin, thanks to the Hanning window.

The last disturbance studied in this work with SK is sag. This is a sudden and temporal amplitude reduction, from a half cycle to a few seconds. The limit is not clearly defined, but when it is raised to a few minutes, it is considered as a change in the power system voltage. This fast change of level in 50 Hz (in the European power system) could be as much as a 90% reduction (more is considered as an interruption of power), but it only occurs temporarily, so the stability indication of this frequency does not change, or almost does not change. However, the amplitude change itself is measured as a leakage in 50 Hz (as seen for other spectral components). Figure 4.20 shows a 10-cycle sag, with a sudden change and a deep of 20% (the change of amplitude in relation to the fundamental one). The start and end are in different realisations.

As leakage occurs in two realisations and amplitude of the main signal that produces this leakage is really large in relation to noise (now the amplitude is much higher than that studied in harmonic examples), the leakage rises to a maximum value related to the impulsive behaviour, when it occurs in two realisations, $N/2$. As 100 realisations are used, this maximum is $100/2 = 50$.

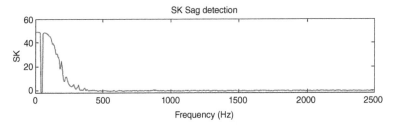

Figure 4.20 SK for a sag disturbance that starts and ends in different realisations. *Source:* Authors.

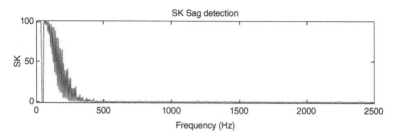

Figure 4.21 SK for a sag disturbance that starts and ends in the same realisations. *Source:* Authors.

Sag disturbance starts and ends in the same realisation and impulsive behaviour could reach a maximum value, N. This can be seen in the following example, when a 20% deep and 5 cycles length sag is simulated, with the start and end in the same realisation. The result of this simulation is shown in Figure 4.21.

Now a similar response can be seen, reaching to the maximum impulsive value (the number of realisations). Even when only a very short part of a sag is inside one realisation and a long part of it is in another, this reaction can be observed, due to leakage of one realisation being higher than in the other.

Oscillations in leakage are related to the position of transitions inside the realisation. Depending on the exact position of the change, a combination is needed, but the average of the oscillations is the same, with the dome related to the spectral leakage. These oscillations imply a non-uniform distribution of leakage, but a higher presence of some frequencies.

5

Measurement Campaigns and Virtual Instruments

5.1 Introduction

The current high rate of penetration of energy resources in the smart grid (SG) has set a challenge for energy management and consequently for supply monitoring. Actually, new types of electrical disturbances have appeared, usually multiple, due to the uncontrolled connection of non-linear loads that can damage the sensitive electronic equipment. For this reason, continuous monitoring of power quality (PQ) is the most effective way to characterise the power system behaviour in order to be able to determine the origin of problems and the cause of a power network fault and to have evidence in reporting that has a bearing on claim damages. Permanent operation supervision will be necessary later because historical observation is the only way to consider what will be the normal or abnormal state on each specific 'site' of the installation under all these multiple influences. Likewise, more holistic indicators that would compute the different network states and deal with big data management in terms of compression and scalability of the measurements are needed in the modern grid.

In this context, this chapter presents a site characterisation through higher-order statistics (HOS). The main objective of the present study is the characterisation of a network node based on a continuous monitoring campaign during different weeks.

PQ monitoring demands more efficient data management, ensuring a flexible reporting, improving the monitoring of different sites and allowing traceability and repeatability in the PQ measures. Moreover, nowadays grid

Power Quality Measurement and Analysis Using Higher-Order Statistics: Understanding HOS Contribution on the Smart(er) Grid, First Edition. Olivia Florencias-Oliveros, Juan-José González-de-la-Rosa, José-María Sierra-Fernández, Manuel-Jesús Espinosa-Gavira, Agustín Agüera-Pérez, and José-Carlos Palomares-Salas.
© 2023 John Wiley & Sons Ltd. Published 2023 by John Wiley & Sons Ltd.

performance analysis implies the characterisation of multiple measurement locations, which results in the production of huge amounts of data that require organisation. This in turn must be based on the elimination of redundant and erroneous information, and on the formulation of new indicators that bring together the greatest possible significance of the measurements that have been adapted to the customer's requirements.

All in all, the reporting levels along with the measurement allocations within the entire network are usually interpreted through the so-called PQ triangle [108]. This graphical representation consists of a data framework in which the concepts of time and space compression are associated with each physical element within the electrical network, along with the magnitudes.

Site analysis and continuous monitoring is needed in the SG scenario. More information needs to be examined in real time while the introduction of a new kind of disturbances that are not characterised today. For example, variations due to wind and solar production will increase on a time scale that is too slow to result in flicker but too fast to be considered as a slow voltage variation (averaging periods of 1 or 10 minutes). Variations in this range are expected to increase in magnitude and at installations where there are several solar panels connected to the same low voltage feeder this may lead to a noticeable increase in voltage variations on this time scale [109]. In this sense, CIGRE-CIRED has recommended that these variations should be further monitored. Suitable PQ characteristics are being developed to quantify the level of variations in a continuous analysis. Finally, monitoring the network continuously would help to develop models that characterise nodes under normal operating conditions in a long-term analysis [110, 111].

In this first section of the chapter, a virtual instrument has been developed based on a National Instrument hardware solution to continuously process the node and acquire the statistics. In the second section, evolution of the statistics in histograms and time series is studied. As a result, several ranges are defined of voltage behaviour under normal conditions and characteristic conditions of that network node.

All the voltage measurements have been done in public networks by the University of Cádiz and household consumers.

5.2 Virtual Instrument

A virtual instrument for PQ assessment based on HOS is proposed in this section. The instrument implements a monitoring strategy that triggers a measurement procedure when an electrical fault is present and a local predefined threshold is surpassed. The method helps to classify events and continuous disturbances by tracking deviations in statistical parameters from their ideal steady-state values.

Designed in LabVIEW™, the user interface includes online graphs showing variance, skewness and kurtosis, along with multiple representations of variance vs. the cited HOS. Based on a 50 Hz 100-signal battery, which gathers different types of electrical disturbances, the instrument was validated during online measurement sessions. Using stat-vs.-stat graphs that implement cycle-to-cycle surveillance, results are depicted in 2D planes. The graphs show that voltage sags and transients could be classified within different clusters, with a low level of uncertainty. These visualisation features allow the operator to view the relevant data objectively during an online monitoring session and, if needed, enhance the report with additional data, for example if they need to claim for economic losses and potential breaches of contract.

The instrument was implemented in the virtual platform LabVIEW, in order to monitor the supply voltage in real time. The instrument measures the signal stability through the PQ index. If a disturbance surpasses the PQ threshold, it triggers the storage unit and a classification analysis takes place within a visual frame (the HOS planes), in order to quantify the degree of the event and its deviation from the ideal contractual power supply.

The HOS plane reports the quality of the signal in terms of its shape and its deviation from the ideal state of the voltage signal. The HOS plane represents these deviations and the PQ index helps quantify them in a single indicator.

Previous studies have proposed the use of cluster analysis. In an earlier study using this instrument [74, 112] the authors focused on the HOS average. The approach used with the current instrument follows the point-to-point behaviour of the recorded deviations and aggregates more relevant information in real-time.

The authors detected that point-to-point tracking allows the discrimination of new zones in the HOS plane related to sag and swell, as demonstrated in reference [74]. These new zones can be associated with signal disturbances, such as sags and transients. In addition, the phase in which the event occurs influences the cycle-to-cycle position in the HOS plane and the trajectory that follows the whole disturbance. These transition zones are addressed in reference [74]. The new cluster zones and their detection can be indicative of the advent of an event or change in the state of a signal suffering a continuous disturbance, such as in the case of sag [74].

5.2.1 Measurement Analysis Framework

The signals studied through the index were acquired directly from the power line at a sampling frequency of 20 kHz. The index was built from data about a 50 Hz sinusoidal voltage signal, sampled using a sliding window that, cycle-to-cycle, computes the HOS estimators. Each cycle comprises 400 time-domain data in a single point in HOS 2D tracking, which is a time-compression strategy to save data storage space and enables coherent organisation of the information about the modern SG.

In order to reduce the computational demands, the 400-point sliding window sweeps the signal cycle-to-cycle without overlapping.

The instrument follows the analysis steps given in the organigram in Figure 5.1 (data acquisition – PQ evaluation – classification analysis).

First, in the data acquisition session the voltage signal was acquired through the virtual instrument. The HOS analysis was based on HOS measurements and the PQ evaluation according to a voltage ideal signal of 50–60 Hz. An ideal signal would obtain a PQ index based an HOS equal to 0. This is an ideal case and all real-life signals contain a certain level of perturbation, thereby producing PQ indices higher than 0.

Second, in order to detect relevant behaviour in the tested signal, a threshold was established by the operator according to the point area being tested and the needs of the monitoring session. The classification analysis was triggered once the PQ index surpassed the specified threshold, indicating unusual signal behaviour. The detection procedure helped generate a database of deviations related to the previously selected threshold. A classification analysis was performed in three steps: HOS representation, cluster analysis and trajectory analysis (point-to-point) (Figure 5.1). The main objective was to classify the recorded deviations as either events or continuous disturbances and to represent them in a more intuitive way.

5.2.2 Experimental Strategy for PQM Through a Virtual Instrument

5.2.2.1 Off-Line Session

As described in the previous section, in order to evaluate HOS plane detection in the instrument, a battery of data on real-life signals, sags and transients was analysed from a building node at the University of Cádiz. The deviations were of different durations (100 signals: 50 sags and 50 transients) at a sampling frequency of 20 kHz (400 samples per cycle). Transient disturbances were detected when the PQ value was superior to 0.02. Sags were detected when the PQ value reached 0.05.

5.2.2.2 Measurements in Field

The characteristic PQ of the monitoring point was established according to the behaviour of the quality of the voltage during the monitoring period. Several data acquisition strategies were used based on the threshold of

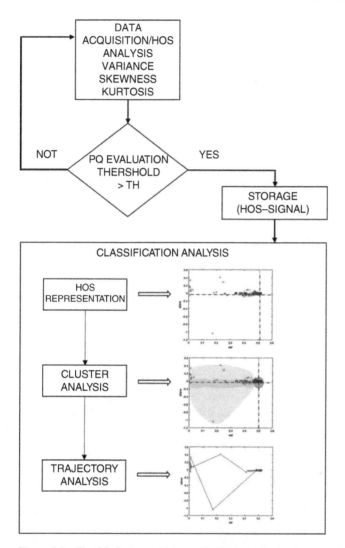

Figure 5.1 The PQ data acquisition procedure and the implementation of a PQ index threshold based on HOS. Classification analysis of the two different strategies: cluster analysis and trajectory analysis of events. *Source:* Olivia Florencias-Oliveros et al. taken from reference [112].

the PQ index. Real-time analysis commonly gave PQ values around 0.026 (an ideal signal would be PQ = 0).

The aim of this was to calibrate the PQ value under normal conditions and set a PQ 'floor'. The test point was set as an average PQ floor value of 0.02. An empirical PQ threshold was derived from analysis (more than twice the value of the previous PQ floor) PQ ≥ 0.05, in order to detect relevant events and study their behaviour in the HOS planes.

5.2.2.3 Study Strategy

It is important to emphasise that these PQ thresholds derived from the instrument are characteristic of the measurement locations being tested. Several testing sessions were carried out prior to using the instrument in an experimental monitoring session, in order to derive suitable thresholds in accordance with the test objectives.

The methodology of the instrument can be used both for analysis of the signal in real time and, in the case of an offline session, in the analysis of a battery of recorded signals.

5.2.3 Configuration of the Virtual Instrument

The instrument configuration follows the organigram and measurement methodology proposed by UNE-EN-61000-4-30 (Figure 5.2). The main objective was to design an instrument (class S) for statistical applications, such as measurement and evaluation of the power supply.

The hardware consists of certificated instruments, such as those by National Instruments, with an Ethernet connection and the NIcDAQ™ 9188 chassis. The analogue input module was the NI-9225 model from the C series. In an earlier version of the instrument, the authors used a Hameg HZ 115 differential probe, in order to amplify the signal from the consumer power line (measuring transducer). The authors decided that the NI9225 module would give

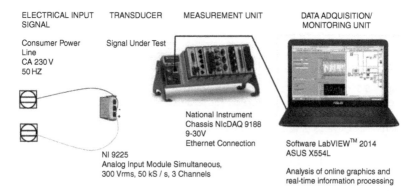

Figure 5.2 The configuration of the virtual instrument guarantees measurement tracking of the devices. *Source:* Olivia Florencias-Oliveros et al. taken from reference [112].

better accuracy when testing the signal. The interface can bring up on-line graphs for analysis, processes signals in real time and report incidents.

All operator level management of monitoring was made possible through a portable computer ASUS X554L via an Ethernet connection. The virtual interface was designed in the LabVIEW (2014) environment so that it could be evaluated before further development and the introduction of improvements related to new quality parameters of the power supply.

The instrument interface used different panels that worked primarily in two PQ control modes (see Figures 5.3 and 5.4).

1) Mode 1. Characterisation of the power line. This mode was aligned with the standard UNE-EN 61000-4-30 [1]. It was used for real-time monitoring of the voltage, sags, swells, transients and harmonics deviations. The measurement procedure continuously analysed windows of 10 cycles (10Tsignal), with a critical time interval measurement of 15 blocks of 10Tsignal (150Tsignal).

2) Mode 2. Detection of transients. This mode used an alternative procedure to the measurement requirements of the standard. It used a sliding window method. The main objective was to reduce the measurement resolution to each cycle and to obtain greater measurement precision.

The main difference between the two mode settings, characterisation and detection, is related to the measurement resolution (basic window length) and their PQ index results. The transient detection mode (Figure 5.3) shows the PQ index in real time, while the characterisation mode averages the measurement window data (PQ values of 10 cycles; 150 cycles).

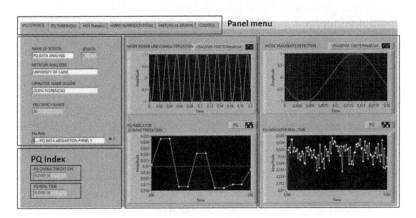

Figure 5.3 The PQ control panel. *Source:* Olivia Florencias-Oliveros et al. taken from reference [112].

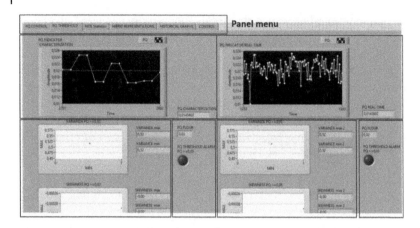

Figure 5.4 The PQ threshold panel allows monitoring of PQ behaviour in both modes for the power line and transients. *Source:* Olivia Florencias-Oliveros et al. taken from reference [112].

The user interface was managed from a main panel, the PQ control (Figure 5.3), so that the operator could configure information related to the working session, including: the network analysed, the operator, the session number and the file path for signal data storage. In Figure 5.3, the control panel, the left side allows the operator to configure the session and save the information. The rest of the panel presents the different operating modes: characterisation of the power line (left) and detection of transients (right), following PQ index trends in both modes. The horizontal line in the PQ panels (grey) represents the PQ floor = 0.02.

This data was used to identify the header of each of the files generated in the database. Additionally, it was possible to track the signal acquisition, the PQ index trend, the PQ characterisation and the PQ in real time. The operator also had the ability to abort the session at any time and the PQ threshold panel (Figure 5.4) could be used to choose a predefined threshold or for the operator to set a threshold manually, according to the session behaviour. The main variables shown in this panel were the PQ value, the HOS disturbance trend, a PQ floor and the PQ threshold alarm.

In Figure 5.4, the operator could select a conventional configuration or a specific one, depending on the variables to be monitored, the thresholds desired and the data to be saved. The transient detection mode is more effective at detecting PQ events and more accurately measures the event evolution. This includes when the PQ floor threshold ($PQ = 0.02$) and the PQ alarm threshold ($PQ \geq 0.05$) are passed.

The input of the instrument automatically computed the state of the signal from the power line, establishing PQ behaviour. The virtual instrument (implemented in LabVIEW) helped to monitor the network and its configuration added flexibility to the working session.

5.2.4 Results

The PQ index evaluation helps the classification analysis that allowed the characterisation and configuration of two real datasets (IEEE DataPort), real PQ sags [108] and real PQ transients [109], available to the scientific community.

The selection of desired limits depends on the behaviour of the signal at the monitoring point. Improved knowledge of this allows for better setting of thresholds according to the needs of the operator. LabVIEW offers multiple possibilities, including: saving graphical representations of PQ behaviour, saving a cluster analysis or the trajectory of a relevant event and updating the online session and other available options shown in the panels (Figures 5.3–5.5). In order to reduce the uncertainty of measurements made by the instrument, instrument design is a key issue. HOS surveillance can be presented through real-time histograms and the standard deviation of individual HOS (Figure 5.5). Also, the statistical range of the monitored variables, and max and min values, can be controlled.

An additional incorporation to the virtual instrument based on HOS is the bi-dimensional planes (2D HOS planes) analysis discussed in the previous chapters. In this case, the representation corresponds to a variance vs.

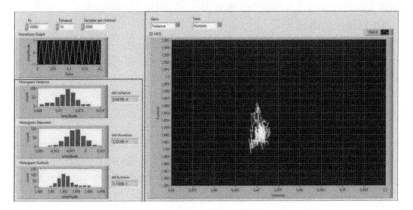

Figure 5.5 The PQ HOS statistics panel. Surveillance of HOS with standard deviation and histograms are on the left. On the right, the operator can observe the analytical trajectory of events within the 2D HOS planes. *Source:* Olivia Florencias-Oliveros et al. taken from reference [112].

skewness, variance vs. kurtosis and kurtosis vs. variance (see Figure 5.5). The instrument includes other graphs such as diverse maximum and minimum graphs corresponding to the real value of individual statistics and depending on the behaviour of a number of data cycles (limits) or a pre-defined interval -th (is an interval selected by the operator) of the analysis: variance (max vs. min), skewness (max vs. min) and kurtosis (max vs. min). Analysis of the limits set offers the possibility of modifying a real session according to empirical experience and the introduction of alternative measurement procedures.

The dynamics of events could be characterised in the HOS planes according to the trajectories of events. Cycle-to-cycle surveillance in the field helps to improve our understanding of the behaviour of events in the HOS planes. More information related to conventional PQ events, such as sag, swell and transients, could be extracted from this new space of representation, analysis and classification (HOS planes).

The quantification of HOS aligned to the events characterised adds more information, allowing us to identify different zones in the HOS planes introducing this information into the PQ advanced applications according to the IEEE Recommended Practice for Monitoring Electric Power Quality–2019 [12].

In the following section, the focus of the virtual instrument is to measure the network continuously and to store data from various weeks with the aim of characterising the node state under normal operating conditions in a wider analysis in MATLAB™.

5.3 PQ Continuous Monitoring Based on HOS for Consumer Characterisation, Public Networks and Households

The main goals of the proposal are described below. HOS are capable of characterising the waveform distortion based on their probability distribution function (PDF) by reporting parameters such as symmetry, amplitude and tailedness of the waveform distribution. Thus, for continuous PQ analysis this section proposes a continuous monitoring solution, as shown in Figure 5.6, through the cycle-to-cycle HOS evolution that helps characterise the waveform in a node of a consumer installation under normal operating conditions. Different measurement scenarios are given in a permanent regime (daily/weekly). This will help in future prediction scenarios.

The solution was implemented as an extension of the real-time monitoring system supported by National Instruments and LabVIEW.

The main objective of this strategy is to learn about the waveform pattern under the contractual operating conditions of the network. Additionally,

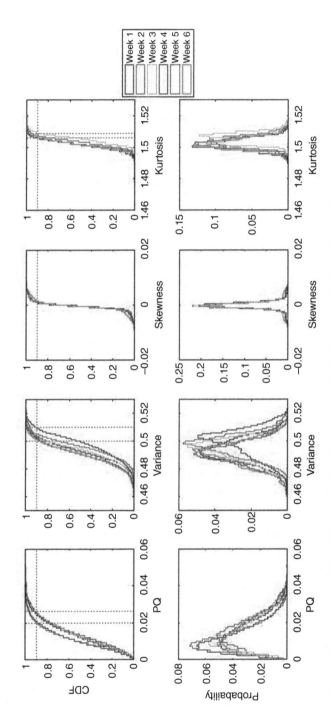

Figure 5.6 Histograms of the different weekly indices based on the CDF and the PDF. *Source:* Olivia Florencias-Oliveros et al. taken and adapted from reference [113].

the information relative to the quality of the waveform helps to simplify the future monitoring campaigns with less computation effort and resources from the beginning. The proposed PQ indicator for the permanent monitoring is to focus on dealing with the large amount of data related to online analysis.

5.3.1 Measurement and Analysis Framework

All the measurements have been made in the installations at the University of Cádiz in Andalusia, Spain. The monitoring campaign has taken a duration of six weeks. The goal was to monitor the 50 Hz voltage waveform in LV at a sampling rate of 20 kHz (400 Sa/cycle). Devices used in the acquisition system are described in the previous section.

In order to stablish a procedure to study the different patterns, all the measurements were analysed off-line a posteriori in MATLAB. The general procedure shown in Figure 5.6 follows the next steps:

1) Extracting HOS and computing PQ cycle-to-cycle in LabVIEW.
2) Store the HOS and the PQ time series in a weekly trend data set.
3) Normalising and pre-processing the HOS and the PQ week time series in MATLAB using the histogram weights. In order to depict a representative percentage of the HOS and the PQ time series in order to extract the PQ trend with the minimum of information, 1 data was selected each 1000 points per bin. Using this technique, authors have reduced the data-set memory storage.
4) Plot the cumulative distribution function (CDF) and the histograms of the HOS and the PQ indicator.
5) Select the HOS and the PQ range based on weekly trend measurements.
6) Plot the PQ time series. Represent the HOS day pattern feature in different colour maps along the weeks.
7) Extract day colour map matrices for PQ pattern detection.

There are two levels of data compression. The first is a temporary compression that takes place cycle-to-cycle and summarises the waveform characteristics in three statistic values (see Table 5.1). The second resides in reducing the histograms of the indices up to the relevant information in each bin (1 data each 1000 points per bin).

5.3.2 Evolution of the Individual Statistics Histograms During Several Weeks

Figure 5.6 shows different histograms of the indices used in the method: the PQ index, the variance, the skewness and the kurtosis. All the measurements were obtained during six weeks from the same connection point under test.

Table 5.1 Site characterisation and data compression through the PQ index for waveform.

Variable	Sizes	Bytes	Class
Waveform	500×1	4000	Double
PQ week	$29\,685\,000 \times 2$	$474\,960\,000$	Double
HOS	1×3	24	Double
PQ	1×1	8	Double

Source: Adapted from Olivia Florencias-Oliveros et al. taken and adapted from reference [113].

The first four boxes in Figure 5.6 (above) represent the cumulative distribution function (CDF) estimate of each index, the PQ and the individual statistics. Figure 5.6 (below) shows the PDFs of each indicator (where it says Probability, it refers to probability density function). In the CDF analysis, each N value is equal to the cumulative relative number of observations in the corresponding bin and all previous bins, that is the way N (end) is equal to 1. In the PDF, each N value is equal to the relative number of observations in the bin such that sum (N) is 1. All the histograms were previously normalised.

In the CDF histogram of PQ, the 90% of the weekly values achieve 0.026 or below and the maximum values reach the range 0.04–0.06. In the histogram corresponding to the PDF, the values of the supply quality index are skewed to the left, which confirms that the trend has a high probability of reaching the value of 0.01. This issue is relevant because the theoretical value for the PQ is zero.

The second and fourth weeks seem to adopt similar trends that are different from the rest of the six weeks. In addition, high values of 0.12 denote the presence of events. However, to achieve a better interpretation of the graphs, a representation range of 0–0.06 has been selected.

The CDF histogram of the variance shows a wider range of variation, with 90% of the values included in the interval [0.49, 0.51], and which includes the theoretical value of the variance, 0.5. A dispersion of 0.01 is therefore permissible. With regard to the PDF histograms of the variance, except for those corresponding to the fourth week, they all exhibit a symmetrical distribution pattern with a very high probability of being within the range [0.49, 0.51].

In relation to the following index, the symmetry or skewness coefficient, the CDF histogram graph confirms that for the power supply voltage signal the skewness denotes a symmetrical behaviour (around the ideal value of zero) for more than 90% of the cycles analysed, with a very small deviation value, and therefore is also admissible.

Finally, and now paying attention to the statistical analysis of kurtosis, with respect to the CDF graph, more than 90% of the values of this index fall below 1.51 (remember that the theoretical value of kurtosis is 1.5). If we pay attention to the PDF chart, the highest values of kurtosis fall within the range [1.5, 1.51], being [1.495, 1.552] for the general range for this index.

Therefore, according to the weekly histogram results, the skewness and kurtosis seem to have the most stable range while the variance shows a wider range. The results obtained, mainly those related to variance, help us to understand that the majority of fluctuations that occur during a week are associated with changes in variance (amplitude changes).

In addition, on the other hand, changes in the tails of the PDFs present a high probability of occurrence, evidenced by the deviation observed in the kurtosis index.

This fact contrasts with the less frequent changes in the symmetry of the distribution, as a working hypothesis for subsequent experiments that finally have the same voltage supply point. However, we must bear in mind that skewness is another term for the PQ index, which will change as specific deviations in the symmetry of the signal will contribute to change the PQ index based in HOS.

5.3.3 Time-Domain Evolution and Bi-dimensional Analysis of the Individual Statistics for Weeks and Days

Next, and in order to obtain daily patterns of 2D graphs, two weeks have been compared based on the variance vs. kurtosis graphs in Figure 5.7. The values with the maximum temporal resolution of 0.02 second, cycle-by-cycle, have been considered. For each day of the considered week, its intensity graph of variance vs. kurtosis (2D colour maps of HOS, named as Fingerprint see Chapter 2, Section 2.8) has been considered, being the skewness obviated because they do not experiment a relevant change under normal operating conditions (more useful in event detection). Similarities have found between patterns of different days within a week and for days corresponding to different weeks.

The measurements took place from Monday to Sunday, from 13 to 28 November 2017. During these two weeks all the cycles of the measured signals were processed except for a lack of 7.2% in week 1 data and a 0.6% in week 2 data (missing 12 hours and 1 hour of monitoring data respectively) according to a loose connection between the acquisition unit and the PC during the monitoring campaign.

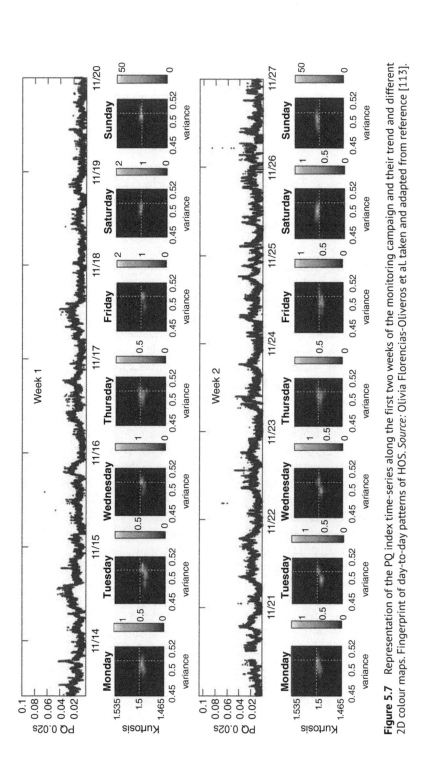

Figure 5.7 Representation of the PQ index time-series along the first two weeks of the monitoring campaign and their trend and different 2D colour maps. Fingerprint of day-to-day patterns of HOS. *Source*: Olivia Florencias-Oliveros et al. taken and adapted from reference [113].

However, in order to carry out a robust characterisation with the least number of data, only a representative part was taken. The criterion adopted was to take data of every 1000 for each container or subdivision made in the histograms of the second compression performed, which actually consists of eliminating redundant information. Before carrying out the compression, the time-series of PQ measurements seems to have coupled noise, which is the visual effect produced by the accumulation of data. Even so, is was observed that the cycle-by-cycle time-series of PQ measurements exhibited a trend that is easily reproducible using a simple mathematical model.

During the first week, there was a lack of PQ data during Monday 13th, because the acquisition started during the noon. Indeed, during nights the PQ index was nearer to the ideal value of 0 and reached the maximum 0.04 in the noon. In addition, there was a second maximum around 0.03, corresponding to the network behaviour during the afternoons. This was observable on Tuesday 14th, Wednesday 15th and Thursday 16th. On Wednesday 15th, some outliers had achieved a PQ of 0.07. In addition, Friday 17th time-series showed that the PQ was lower during the afternoon, similar to the night periods. Moreover, Saturday 19th and Sunday 20th exhibited a completely different PQ pattern with a general PQ trend concentrated in the interval [0.01, 0.02].

Let us now focus on the colour maps corresponding to one day fingerprint of HOS, which represent the variance vs. kurtosis. They show regions with shades close to white. As was the case in histograms, the variance reached the greatest elongations, which are represented in the pattern of the graph along the horizontal axis. For some days, two centroids were observed in the 2D graph that correspond to two different states of the network during the same day. This fact is even more visible on Tuesday 14th, Wednesday 15th, and Thursday 16th. On Saturday 19th and Sunday 20th, the pattern is more diffuse at the periphery of the graph and is more intense in the centre. Precisely the greatest number of measurements occur on Sunday 20th, nearest to the point that corresponds to the ideal supply value.

Furthermore, during the second week, the PQ time series follows a similar behaviour to that in week 1. There is a lack of PQ data during Monday 13th in the noon as result of the monitoring campaign. From Monday 21st to Friday 25th, two maximum regions during the day can be detected and a lower behaviour was seen during the nights around 0.01–0.02. A more unstable pattern can be seen on the weekend if compared with the same period during week 1. However, in general terms at the weekends the PQ trend is, most of the time, near to the ideal values. Some outliers can be observed on Wednesday 23th and Sunday 27th that reach a PQ of 0.08 and 0.1 respectively.

Besides that, day-to-day colour maps confirm the patterns in two different clusters from Monday to Friday and the weekends with the most centre behaviour according to the PQ trend. The minimum PQ value is always over zero, a fact that is indicative of a non-ideal behaviour of the voltage supply waveform under normal operating conditions.

5.3.4 Data Organisation

The CDF and PDF analysis helps to stablish a characterisation analysis of the waveshape of the point under test. For a weekly/monthly campaign, the indices most representative seem to be variance and kurtosis. Nevertheless, it is important to say that, based on author's experience, skewness can be useful in strategies more focused on small length windows and event detection.

The main contribution to the histogram's analysis is to set more realistic measurements to the individual ranges of indices and detect their region of maximum probability within the whole campaign.

The time domain analysis helps to detect the effect of the waveform deviation computed by the statistics. It also allows identification of the individual contribution of each indicator to the global PQ. Indeed, in the network under test, the PQ day fluctuates depending on the day of the week, the hourly trend of the network and the energy usage during working or non-working hours. During the night, in general the PQ cycle-to-cycle fluctuates between 0.01 and 0.02 values. During the morning, the deviation of the PQ increases, reaching maximum 0.04 values by the noon. In addition, between 13 : 00 and 16 : 00 hours there is a drop in the PQ because this is the lunch schedule in Spain. A second increase of the index occurs in the evening schedule, since the University is open until 22 : 00 hours. Finally, during midnight the PQ decreases again, recovering the minimum values. Indeed, 2D graphics based on fingerprint in the HOS planes allow visualising such behaviour by emphasising the areas of signal persistence throughout the hours, days and weeks.

In order to make a site characterisation through the PQ features, the time-series can be scaled to different average windows. For example, Figure 5.8 shows the 10 minutes and hourly PQ trend for the 2 weeks that had previously been analysed in Figure 5.7. Notice that the average of the PQ values help to compress the time-series and extract the information about the general trend of the index in the network to a minimum computation report. Regarding the permanent monitoring strategy, the hourly PQ mean is used to produce the PQ cycle-to-cycle trend chart. To set a compliance limit of the values, following the average PQ strategy, they must be 0.03 during the 95% (Figure 5.8). Nevertheless, notice that, by averaging the indicator, the

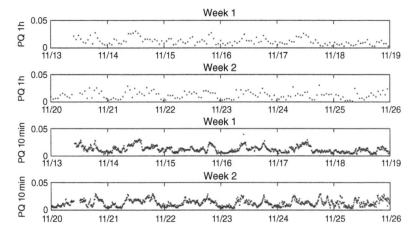

Figure 5.8 Different PQ monitoring strategies informing about the PQ mean each hour and the PQ mean each 10 minutes for two previous weeks. *Source:* Olivia Florencias-Oliveros et al. taken and adapted from reference [113].

PQ loses the event detection resolution but the measurements help to train the algorithm to provide a more accurate PQ control with less information.

As a general remark, the PQ trend values can variate from one network to another and depend on the quality of the voltage waveform shape in the point under test, which can be influenced by multiple factors, including changes in voltage admitted by the actual PQ standards. Depending on the customer and network requirements, this non-ideal behaviour may or may not be assumed.

In general, both the utility and the end-users would facilitate that information, introducing the continuous PQ monitoring of the waveform shape and characterising their deviation under normal operating conditions. In this sense, the minimum requirements in terms of computation are summarised in Table 5.2.

Table 5.2 Site characterisation and data compression through the PQ index for a continuous monitoring campaign.

Variable	Sizes	Bytes	Class
HOS week	29 685 000 × 6	1 424 880 000	Double
PQ week	29 685 000 × 2	474 960 000	Double
Matrix vs. k	102 × 102	83 232	Double
PQ 1 h	164 × 2	2624	Double
PQ 10 min	990 × 2	15 824	Double

Source: Adapted from Olivia Florencias-Oliveros et al. taken and adapted from reference [113].

Reporting the PQ every 10 minutes in a month would need ($15\,824 \times 30$ days) $474\,720$ bytes and in a year ($474\,720 \times 12$ months) $5\,696\,640$ bytes, which is equal to 0.005 696 64 gigabytes. The PQ strategy deployed in a thousand smart meters would demand a storage capacity of 5.69 gigabytes in a year. All in all, this demonstrates that the procedure is more than appropriate to be incorporated in future PQ measurement campaigns and that the solution would accomplish monitoring challenges of the next generation of advance metering infrastructure in terms of compression and reporting PQ efficiently.

5.3.5 PQ and Energy Patterns as a Scalable Proposal

By analysing the time series of PQ more closely, certain patterns can be detected in order to characterise different consumer behaviours and their connected loads. In this section, different pattern detection strategies when measuring in the field will be analysed according to proposals previously introduced.

Modelling a PQ parameter such as voltage and current in low-voltage networks is only possible using a probabilistic model. Nevertheless, modelling PQ parameters can be difficult because single loads and the mixed effect of a group of loads need to be summarised using probabilistic methods to calculate the total distortion in a PCC (point of common coupling). Nevertheless, modelling a consumer installation is not realistic at all. Indeed, in order to develop a reproducible model of the network state, an advance consumer characterisation under real measurements is required. In this sense, real-time measurements can help to stablish a correlation between the quality of the waveform and their deviations based on the density functions (PDF).

The changes in the waveform distribution can have an origin in multiple unknown circumstances. These would come through connected loads that depend on the behaviour of a certain user or a consumer who uses that load, for example one who turns on the air conditioner during summer or a heater in winter. In the end, the information contained within the consumer behaviour can simply be known by measuring in advance, because they demand multiple reasons, such as the weather, daytime, etc.

As demonstrated in Section 5.3.3, different day patterns can be detected from the site characterisation through the PQ deviation index based on HOS, where the total voltage disturbance level on the monitored line was detected through the index. In this sense, different distributions can be extracted from a continuous monitoring analysis, computing the random variable in a predefined time interval (e.g. a day, an hour). Different days have been selected from the PQ time-series in Section 5.3.3 for weeks 1 and 2.

5.3.5.1 Site Characterisation Between Days

This section looks at the PQ analysis, from which it is seen that, depending on the window length, the focus changes from reliability (wide windows) to PQ (shorter windows). Figure 5.9 shows the 2D planes corresponding to the whole week, from which two relevant days, the best and the worst, have been selected to compare the trace evolution from the one-day frame to the 0.02-seconds frame (equivalent to 10 cycles). The first column corresponds to the one-week results; then, the trace evolution from one day to 0.2 seconds is shown.

The 3×3 2D-HOS graph matrix corresponding to the best (worst) day is located below (above). Within the worst (best) day, the worst (best) hour and the worst (best) 0.02 seconds are selected. Indeed, Figure 5.9 shows details of the PQ temporal organisation. This tree structure can be easily followed from left to right as the temporal width decreases, from reliability (one week) to PQ indications (0.2 seconds monitored). Thus, it can be said that reliability patterns (long width time windows) can be confirmed as the aggregation of different PQ states (shorter windows). Therefore, as the monitoring scale decreases, the focus goes from the concept of reliability to the PQ event detection strategy. This fact illustrates one of the contributions of this measurement strategy, where several analyses can be performed in a unified framework, allowing post comparisons between different time-scales.

For instance, from the pattern in the one-week monitoring period, the worst and the best day can be distinguished as the top and bottom 3×3 matrix block of 2D-HOS graphs, which exhibit extremely different behaviours. The general week pattern resulting from the aggregation of shorter-time states allows the whole PQ history (from a week to 0.2 seconds) to be explored, following multiple sequences of frames and also enabling direct comparisons between frames of the same or different time scales. This provides information regarding deviations experimented in raw power supply data in the surveyed intervals.

Quantitatively, for the worst day (Figure 5.9), the PQ increased as the time window was shortened. Paying attention to the heart of the figures, this shows that the HOS plane allowed details to be extracted from the voltage supplied.

5.3.5.2 Performance Analysis

Figure 5.10 shows the non-stationary evolution of the PQ index within a single day, with an average value of PQ day $= 0.0138$. As a general remark, this shows that the PQ depends on the period of the day.

Hereinafter, the analysis of the top PQ series in Figure 5.10 begins. First, a general analysis regarding the trend of the curve was performed. Between 00 : 00 and 06 : 00, oscillations occurred just above the ideal value of zero, with

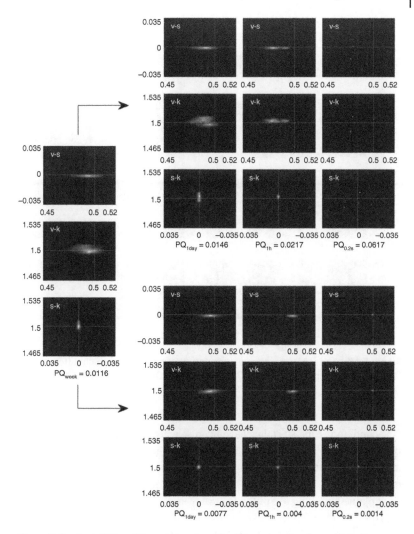

Figure 5.9 Scalability of the proposed method: two singular days within the analysed week. *Source:* Olivia Florencias-Oliveros et al. taken from reference [75].

an average of PQ6h = 0.0094. Then, in the interval 06 : 00 to 10 : 00, the average value increased with the electricity demand, PQ4h = 0.0148. Within the time stretch 10 : 00 to 15 : 00, the quality became worse with a PQ5h = 0.0205, which were the central hours of the observed day. Then, in the interval 15H00 to 20H00, a soft decrease took place as a direct consequence of activity ceasing. The same was observed in the period 20H00 to 24 : 00.

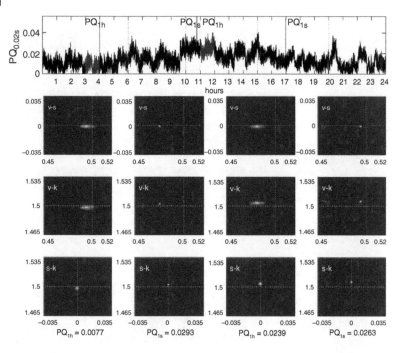

Figure 5.10 The PQ 0.02 second values during 24 hours, from which different intervals can be analysed. *Source:* Olivia Florencias-Oliveros et al. taken from reference [75].

Second, four time stretches have been selected to illustrate more specific behaviours: two relevant hours and two relevant seconds. For each one, its associated triplet of 2D-HOS graphs is drawn just below. The two-hour periods (03 : 04 and 11 : 12) enable the dynamic of the trace through the day to be explained, making use of the PQ time-series and the 2D HOS planes. The PQ1h was very different between the selected hours: first PQ1h = 0.0073 vs. second PQ1h = 0.0237 respectively, with more stability (near the ideal zero value) in the first and worse (above zero) in the second hour under test.

The maximum PQ 0.02 second value was reached at PQ 0.02 second = 0.05, within the analysed second, between 10 : 00 and 11 : 00. This maximum may indicate the presence of at least one event. Thus, the statistical parameters deviated from the ideal values, even out of the trend within this time interval. A singular event was detected in the HOS plane and was part of a series of events that marked a high PQ trend.

The same maximum was reached within the second interval 15 : 00 to 20 : 00. This time, it seemed as if the PQ evolved closer to the ideal value. This is observed in the v–k graph, where two clouds can be differentiated;

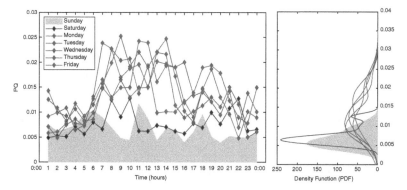

Figure 5.11 Hourly trend of the index and different PQ density functions during several days in the whole week. *Source:* Olivia Florencias-Oliveros taken from reference [69].

the one on the left is associated with the event group and the second one is nearer the origin, where the ideal values are located. The demand intervals 06 : 00 to 14 : 00 and 15 : 00 to 20 : 00 can be extracted from the HOS graphs. They are also reflected in a 24-hour monitoring solution based on the PQ index in Figure 5.11.

Finally, in order to show a strategy extracted from the whole of the raw data, Figure 5.11 represents the hourly trend of the index during several days in the whole week. It is seen that, while a non-working day had a PQ value always near zero, the five working days exhibited a quite different behaviour, reaching maximum values.

Hourly PQ measurements and their PDF helps to extract two-different density functions within the working day and the non-working day data, which show the stationary intervals of the index throughout the week (Figure 5.11).

Working days: Exhibit a symmetric and flattened distribution that comes with higher values of PQ.

Non-working days: With a more narrow and skewed left distribution, indicates the lowest deviation of the index along the day.

5.4 Simplified Method to Use HOS in a Continuous Monitoring Campaign

Figure 5.12 summarises a flowchart strategy to use HOS in a continuous monitoring campaign, combining the information that comes from the PQ index and the 2D HOS planes. Additionally, as has been explained before, some other variables are monitored and correlated, such as the demanded power.

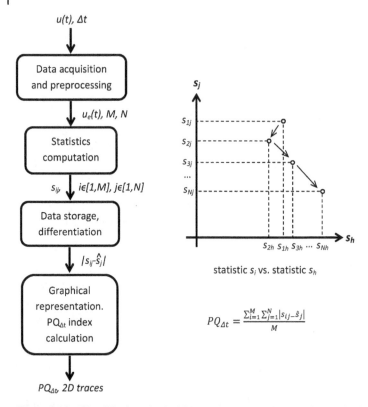

Figure 5.12 Simplified method of PQ continuous monitoring. *Source:* Authors taken from [69].

Findings: The results allow us to conclude that HOS are capable of establishing a continuous, reliable performance that characterises the deviations of the voltage supply waveform in an average consumer installation. The method opens the door to a more feasible solution for PQ surveillance in the SG.

Recommendations: The continuous monitoring solution helps to optimise the statistical studies of HOS on the PQ monitoring field. The voltage information is aligned with the HOS planes and PQ measurements in the EEE Std 1159-2019 [12].

The method could be turned over to data management platforms such as SCADA systems to control the applications at different security levels, optimising different screens and data. Figure 5.13 summarises the simplified way to represent the PQ information within the SG. The procedure has been conceived to obtain a better management of data through compression in time and space (see Figures 5.13 and 5.14. PQ devices such as instruments

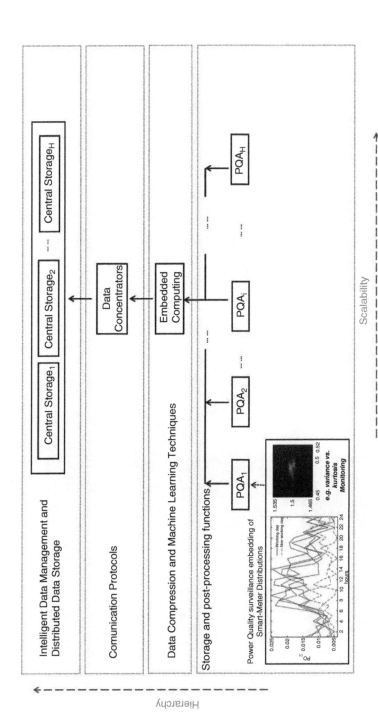

Figure 5.13 HOS continuous monitoring-based procedure in the SG. *Source:* Authors.

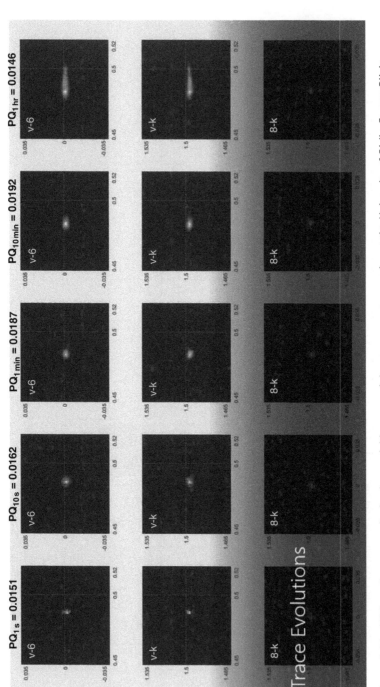

Figure 5.14 HOS continuous monitoring in a scalable proposal. A measurement campaign at the University of Cádiz. *Source:* Olivia Florencias-Oliveros et al. taken from reference [114].

and smart meters with PQ functionalities are responsible for analysing and storage of PQ information at a low voltage level. Based on PQ recorded information, embedded computing at this level provides data compression and machine learning techniques. Data concentrators deal with different communication protocols that are used to transfer data to central storage. Finally, intelligent data management and distributed data storage take place.

5.5 Conclusions

This work has described a virtual instrument that implements an HOS-based global, scalable PQ index, associated with a specific location in the tested network. It was tested and validated using a real data set. The index has a twofold purpose: to set a threshold for the power supply signal and to trigger a PQ analysis once a threshold has been breached. The experimental strategy used real-life signals to demonstrate the validity of the procedure, which assessed the stability of the signal in the time domain through cycle-to-cycle tracking of the time-series of the index.

This HOS-based approach helps in monitoring the PQ dynamics and tracks voltage supply stability in 2D. This also allows us to infer the type of disturbance according to the trajectory displayed in the HOS graphs.

Another concept that emerges from this work is the PQ floor, from which it is possible to fix a PQ threshold associated with a measurement location during a measurement interval. The threshold should be modified according to customer and system requirements. During a monitoring session, the index dynamics enables reporting of PQ trends and helps discrimination by network type and detected deviations, as well as allowing a forecasting analysis.

The classification analysis streamlines the capture process and reduces the storage space required for the data as fewer samples are needed to monitor the voltage signal. It obtains effective results when compared to traditional methods of analysis, such as signal compression.

The instrument has different operation procedures that can be adapted for users depending on their knowledge of PQ. The transient detection mode measures power line disturbances with greater specificity.

Outcomes of the online monitoring sessions confirmed the theoretical approach of clustering found in the different graphs analysed.

A signal database was generated in the IEEE data port related to measurements in the field [115, 116].

The method is based on continuous monitoring and enables a more efficient management of voltage through the PQ index based on HOS. It helps to monitor the network qualitatively through a supervisory control, achieve the process historic, allows training of instruments and compares between different monitoring campaigns.

The method for continuous measurements is optimised using the PQ indicator and the probability density graphs are obtained based on weekly patterns and most significant days and hours.

It has been shown that the solution can be scalable, allowing not only to characterise events or small periods but also several weeks of monitoring campaigns.

- Within a week, trends are reported: working vs. non-working day, from Monday to Thursday vs. Friday and the weekend, which seems to exhibit other patterns influenced by the use of the network. Also, the afternoon is better to connect a sensible load or device.

The method helps to describe the relationships between the PQ and energy demand and the efficiency of the installation.

Appendix A

Voltage Waveform

A.1 Theoretical Power System Waveform

The theoretical power system waveform is a sinusoidal time function $v(t)$, with an ideal constant frequency and amplitude. The theoretical behaviour of the network must be a voltage source without internal impedance, $Z(\Omega) = 0$, is given by a voltage source with a sinusoidal output and constant voltage at all points of supply:

$$v(t) = V_{max} * sin(2\pi ft + \theta);$$

where
ω is the angular frequency $2\pi f$ (rad/s)
f is the power system frequency (Hz)
θ is the phase angle between the current and the voltage (rad)
t is the time (s)

An industrial power system consists of a three-phase feeder, each of the phases being displaced 120° with respect to each other. The network impedance is not zero, $Z(\Omega) \neq 0$.

Power Quality Measurement and Analysis Using Higher-Order Statistics: Understanding HOS Contribution on the Smart(er) Grid, First Edition. Olivia Florencias-Oliveros, Juan-José González-de-la-Rosa, José-María Sierra-Fernández, Manuel-Jesús Espinosa-Gavira, Agustín Agüera-Pérez, and José-Carlos Palomares-Salas.
© 2023 John Wiley & Sons Ltd. Published 2023 by John Wiley & Sons Ltd.

Appendix B

Time-Domain Cumulants

Higher-order statistics estimates have been proposed through the last decade to infer new statistical characteristics associated with data from a non-Gaussian time-series in a predominant Gaussian background floor, which can be theoretically considered as a result of the summation of different noise processes.

Within the PQ detection context, the targeted electrical disturbance is always considered to be non-Gaussian, while the floor is assumed to be a stationary Gaussian signal. Thus, using HOS would help locate the perturbation qualitatively, with a rough approximation, but allowing it to be allocated in a permanent measurement campaign.

Given an rth-order real-valued stationary random process (an original analogue time-series), $x(t)$, we can define a set of random variables (time-series) given by

$$x(t), x(t+\tau_1), x(t+\tau_2), \ldots, x(t+\tau_{(r-1)})$$

(B.1)

where the discretized time is $t = k \cdot T_s$ (k, integer) and $\tau_i = i \cdot T_s$ is the ith time shift applied to each of the resulting time-series from the original measurement data, which in turn is also a multiple of the data acquisition sampling period, T_s. The joint (compacted notation) rth-order cumulant of the random variables is given by the mathematical expression

$$C_{r,x}(\tau_1, \tau_2, \ldots, \tau_r) \equiv Cum\left[x(t), x(t+\tau_1), x(t+\tau_2), \ldots, x(t+\tau_{(r-1)})\right]$$
$$= E\{x(t) \cdot x(t+\tau_1) x(t+\tau_2), \ldots, x(t+\tau_{r-1})\}$$

(B.2)

Power Quality Measurement and Analysis Using Higher-Order Statistics: Understanding HOS Contribution on the Smart(er) Grid, First Edition. Olivia Florencias-Oliveros, Juan-José González-de-la-Rosa, José-María Sierra-Fernández, Manuel-Jesús Espinosa-Gavira, Agustín Agüera-Pérez, and José-Carlos Palomares-Salas. © 2023 John Wiley & Sons Ltd. Published 2023 by John Wiley & Sons Ltd.

From Equation (B.2) it can be seen that the cumulant is defined as the autocorrelation product between the original time-series and their associated time-shifted versions. The objective of the cumulants is to quantify the mathematical similitude in order to infer potential statistical interdependency. Depending on the product order, the interdependency will lead to specific behaviour within the system from which the measurements have been extracted. The most common cases of the cumulants are the second-, third- and fourth-order versions, defined as follows:

$$C_{2,x}(\tau) \equiv E\big[x(t)\cdot x(t+\tau)\big] \tag{B.3}$$

$$C_{3,x}(\tau_1,\tau_2) \equiv E\big[x(t)\cdot x(t+\tau_1)\cdot x(t+\tau_2)\big] \tag{B.4}$$

$$C_{4,x}(\tau_1,\tau_2,\tau_3) \equiv E\big[x(t)\cdot x(t+\tau_1)\cdot x(t+\tau_2)\cdot x(t+\tau_3)\big]$$

$$-C_{2,x}(\tau_1)\cdot C_{2,x}(\tau_2-\tau_3)$$

$$-C_{2,x}(\tau_2)\cdot C_{2,x}(\tau_3-\tau_1)$$

$$-C_{2,x}(\tau_3)\cdot C_{2,x}(\tau_1-\tau_2) \tag{B.5}$$

In the case of a non-zero mean process, $x(t)$ is replaced by $x(t) - E[x(t)]$ in Equations (B.2) and (B.3). If there is no time shifting ($\tau_i = 0$, i), three very well-known measurements of a statistical distribution are obtained:

$$\gamma_{2,x} \equiv C_{2,x}(0) = E\big[x^2(t)\big] \tag{B.6}$$

$$\gamma_{3,x} \equiv C_{3,x}(0,0) = E\big[x^3(t)\big] \tag{B.7}$$

$$\gamma_{4,x} \equiv C_{4,x}(0,0,0) = E\big[x^4(t)\big] - 3\big(\gamma_{2,x}\big)^2 \tag{B.8}$$

The cumulants in Equations (B.6) to (B.8) are normalized in order to obtain three very well-known measurements of a statistical distribution: the variance, the skewness and the kurtosis respectively, according to the following equations:

$$\mathrm{var}(x) = \gamma_{2,x} = E\big[x^2(t)\big] = \mu_2 = \sigma^2 \tag{B.9}$$

$$\mathrm{skew}(x) = \gamma_{3,x}\big/\big(\gamma_{2,x}\big)^{3/2} = E\big[x^3(t)\big]\big/\big(\gamma_{2,x}\big)^{3/2} = \mu_3/\sigma^3 \tag{B.10}$$

$$\mathrm{kur}(x) = C_{4,x}\big/\big(C_{2,x}\big)^2 = E\big[x^4(t)\big]\big/\big(\gamma_{2,x}\big)^2 - 3 = \mu_4/\sigma^4 - 3 \tag{B.11}$$

The ensemble of Equations (B.9) to (B.11) constitutes indirect measurements of the variance, skewness and kurtosis, which are the most well-known statistics for a continuous random variable. If $x(t)$ is symmetrically distributed, its skewness is zero (but not vice versa, which are improbable situations); if $x(t)$ is Gaussian distributed, its kurtosis is necessarily zero (but not vice versa). This standardisation (statistical normalisation) makes estimators shift and scale invariant.

Appendix C

HOS Range for Sag Detection, 1 Cycle

Table C.1 HOS range for sag detection without phase-angle jump based on non-symmetrical and non-sinusoidal conditions, in one cycle.

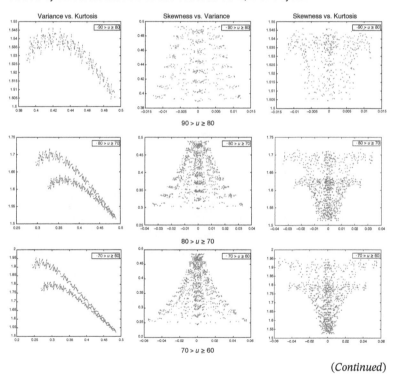

(Continued)

Power Quality Measurement and Analysis Using Higher-Order Statistics: Understanding HOS Contribution on the Smart(er) Grid, First Edition. Olivia Florencias-Oliveros, Juan-José González-de-la-Rosa, José-María Sierra-Fernández, Manuel-Jesús Espinosa-Gavira, Agustín Agüera-Pérez, and José-Carlos Palomares-Salas.

Table C.1 (Continued)

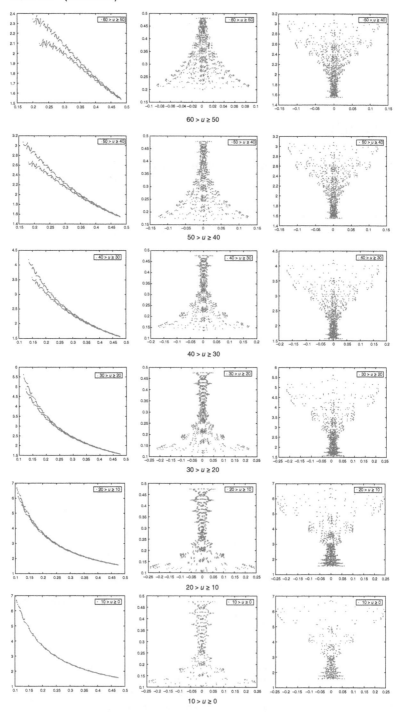

HOS evolution according to cycle-to-cycle amplitude changes and the reference voltage $u\%$ of U_{ref}.

Appendix D

HOS Range for Sag Detection, 10 Cycles

Table D.1 HOS range for sag detection including phase-angle jumps based on non-symmetrical and non-sinusoidal conditions in a simulation of 10 cycles.

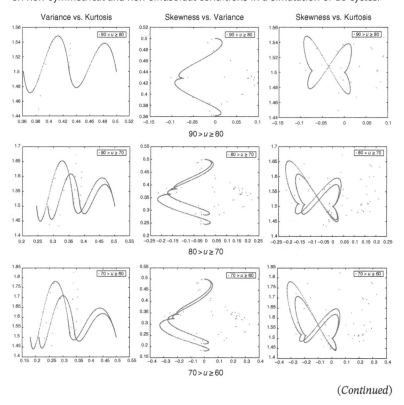

(Continued)

Power Quality Measurement and Analysis Using Higher-Order Statistics: Understanding HOS Contribution on the Smart(er) Grid, First Edition. Olivia Florencias-Oliveros, Juan-José González-de-la-Rosa, José-María Sierra-Fernández, Manuel-Jesús Espinosa-Gavira, Agustín Agüera-Pérez, and José-Carlos Palomares-Salas. © 2023 John Wiley & Sons Ltd. Published 2023 by John Wiley & Sons Ltd.

Table D.1 (Continued)

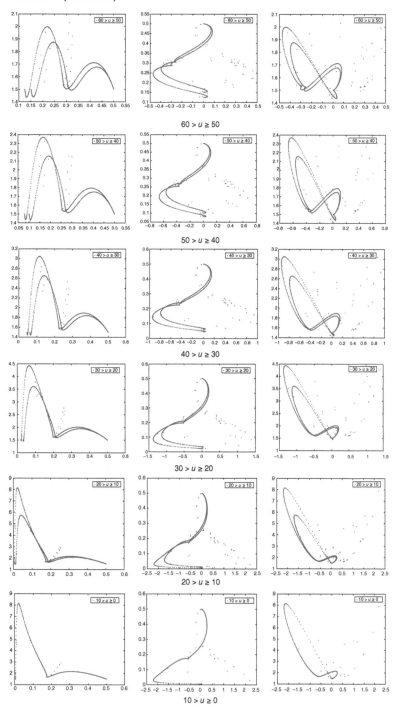

HOS phase angle jump according to cycle-to-cycle amplitude changes and the reference voltage $u\%$ of U_{ref}.

References

1 International Electrotechnical Commission (2015). IEC. 61000-4-30:2015. Electromagnetic compatibility (EMC). Part 4–30: Testing and measurement techniques. Power quality measurement methods. https://webstore.iec.ch/publication/21844 (accessed 20 October 2022).

2 CEN/CENELEC/ETSI Joint Working Group on Standards for Smart Grids (2012). CEN-CENELEC-ETSI Smart Grid Coordination Group: Smart Grid Information Security, 1–107. https://www.cencenelec.eu/media/CEN-CENELEC/AreasOfWork/CEN-CENELEC_Topics/Smart Grids and Meters/Smart Grids/security_smartgrids.pdf.

3 Flores, R. (2003). Signal processing tools for power quality event classification. Licensed English Thesis. School of Electronic Engineering, Chalmers University of Technology, Göteborg, Sweden.

4 Meyer, J., Klatt, M., and Schegner, P. (2011). Power quality challenges in future distribution networks. *2011 2nd IEEE PES International Conference Exhibition of Innovative Smart Grid Technology*, Manchester, UK (5–7 December 2011), 1–6. Institute of Electrical and Electronics Engineers (IEEE). http://toc.proceedings.com/14237webtoc.pdf (accessed 20 October 2022).

5 Muscas, C. (2010). Power quality monitoring in modern electric distribution systems. *IEEE Instrumentation and Measurement Magazine* 13 (5): 19–27.

6 Emanuel, A.E., Langella, R., and Testa, A. (2012). Power definitions for circuits with nonlinear and unbalanced loads – The IEEE standard 1459–2010. *IEEE Power Energy Society General Meeting*, San Diego, CA, (22–26 July 2012), 1–6. https://doi.org/10.1109/PESGM.2012.6345330.

7 Eigeles Emanuel, A. (1990). Power in non-sinusoidal situations. A review of definitions and physical meaning. *IEEE Transactions on Power Delivery* 5 (3): 1377–1389.

Power Quality Measurement and Analysis Using Higher-Order Statistics: Understanding HOS Contribution on the Smart(er) Grid, First Edition. Olivia Florencias-Oliveros, Juan-José González-de-la-Rosa, José-María Sierra-Fernández, Manuel-Jesús Espinosa-Gavira, Agustín Agüera-Pérez, and José-Carlos Palomares-Salas.
© 2023 John Wiley & Sons Ltd. Published 2023 by John Wiley & Sons Ltd.

8 Czarnecki, L.S. (1987). What is wrong with the Budeanu concept of reactive and distortion power and why it should be abandoned. *IEEE Transactions on Instrumentation and Measurement* IM–36 (3): 834–837.

9 Martzloff, F. and Gruzs, T.M. (1988). Power quality site surveys: facts, fiction, and fallacies. *IEEE Transactions on Industry Applications* 24 (6): 1005–1018.

10 IEEE 1459 (2010). PE/PSIM – Power System Instrumentation and Measurements. *IEEE Standard Definitions for the Measurement of Electric Power Quantities Under Sinusoidal, Nonsinusoidal, Balanced, or Unbalanced Conditions.* Institute of Electrical and Electronics Engineers (IEEE). https:// standards.ieee.org/ieee/1459/4088/ (accessed 20 October 2022).

11 Ferrero, A. and Superti-Furga, G. (1991). A new approach to the definition of power components. *IEEE Transactions on Instrumentation and Measurement* 40 (3): 568–577.

12 IEEE Standards Coordinating Committee 22 on Power Quality (2019). *IEEE Recommended Practice for Monitoring Electric Power Quality.* https://doi.org/10.1109/IEEESTD.2019.8796486.

13 European Standards (2015). EN 50160 2011/A1: 2015. Voltage characteristics of electricity supplied by public electricity networks. https://www.en-standard. eu/search/?q=EN+50160+2011/A1:+2015 (accessed 20 October 2022).

14 Milanović, J.V., Meyer, J., Ball, R.F. et al. (2014). International industry practice on power-quality monitoring. *IEEE Transactions on Power Delivery* 29 (2): 934–941.

15 Alsayyed Ahmad, B., ElSheikh, H. H., and Fadoun, A. (2015). Review of power quality monitoring systems. *2015 International Conference on Industrial Engineering Operational Managagement*, Dubai, UAE (3–5 March 2015), 1–8.

16 Zhou, K., Fu, C., and Yang, S. (2016). Big data driven smart energy management: from big data to big insights. *Renewable and Sustainable Energy Reviews* 56: 215–225.

17 Farhangi, G. (2010). The path of the smart grid. *IEEE Power and Energy Magazine* 8 (1): 18–28.

18 Bollen, M.H.J., Bahramirad, S., and Khodaei, A. (2014). Is there a place for power quality in the smart grid? *Proceedings of the International Conference on Harmonic Quality Power, ICHQP*, Bucharest, Romania (25–28 May 2014), pp. 713–717. Institute of Electrical and Electronics Engineers (IEEE).

19 European Commission, Directorate-General for Research and Innovation (2016). European Smart Grids Technology Platform: Vision and strategy for Europe's electricity networks of the future. Publications Office http:// www.manufuture.org (accessed 04 April 2012).

20 Rönnberg, S. and Bollen, M. (2016). Power quality issues in the electric power system of the future. *The Electricity Journal* 29 (10): 49–61. https://doi.org/10.1016/j.tej.2016.11.006.

21 McDermott, T.E. and Dugan, R.C. (2003). PQ, reliability and DG – measuring distributed generation's impact on reliability and power quality. *IEEE Industry Applications Magazine* 9 (5): 17–23. https://doi.org/10.1109/MIA.2003.1227867.

22 Zavoda, F. (2008). The key role of intelligent electronic devices (lED) in advanced distribution automation (ADA). *CICED2008. Technical Session 3. Protection, Control, Communication and Automation of Distribution Network*, Guangzhou, China (10–13 December 2008), 1–7. Institute of Electrical and Electronics Engineers (IEEE).

23 Zavoda, F., Rodriguez, E., and Fofeldea, G. (2017). Underground and overhead monitoring systems for MV distribution networks. *CIRED – Open Access Proceedings Journal* 2017 (1): 477–480. https://doi.org/10.1049/oap-cired.2017.1130.

24 Zavoda, F., Fofeldea, G., and Rodriguez, E. (2019). Underground distribution network monitoring so much easier. *25th International Conference on Electricity Distribution (CIRED 2019)* (June 2019), 3–6. https://www.cired-repository.org/handle/20.500.12455/358. (accessed 04 April 2006).

25 Tcheou, M.P., Lovisolo, L., Ribeiro, M.V. et al. (2014). The compression of electric signal waveforms for smart grids: state of the art and future trends. *IEEE Transactions on Smart Grid* 5 (1): 291–302. https://doi.org/10.1109/TSG.2013.2293957.

26 Ferreira, D.D., De Seixas, J.M., Cerqueira, A.S. et al. (2015). A new power quality deviation index based on principal curves. *Electric Power Systems Research* 125: 8–14. https://doi.org/10.1016/j.epsr.2015.03.019.

27 Meyer, J., Canada, I., Bahramirad, S., and Langella, R. (2015). Paper 0183, Volt-VAR control and power quality (CIGRE/CIRED C4.24), solar power future voltage control. *23rd International Conference on Electricity*, Lyon, France (15–18 June), 15–18. http://cired.net/publications/cired2015/papers/CIRED2015_0183_final.pdf (accessed 10 October 2022).

28 Hyndman, R.J., Liu, X.A., and Pinson, P. (2018). Visualizing big energy data: solutions for this crucial component of data analysis. *IEEE Power and Energy Magazine* 16 (3): 18–25.

29 Wang, Y., Chen, Q., Hong, T., and Kang, C. (2019). Review of smart meter data analytics: applications, methodologies, and challenges. *IEEE Transactions on Smart Grid* 10 (3): 3125–3148.

30 Pereira, M., Bessa, R.J., and Gouveia, C. (2019). Low voltage grid data visualisation with a frame representation and cognitive architecture. *2019 IEEE Milan PowerTech*, 1–6.

31 Beleiu, H.G., Beleiu, I.N., Pavel, S.G., and Darab, C.P. (2018). Management of power quality issues from an economic point of view. *Sustainability* 10 (7): 1–16.

32 CEER (2018). CEER Benchmarking Report 6.1 on the Continuity of Electricity and Gas Supply, p. 84.

33 Elphick, S., Ciufo, P., Smith, V., and Perera, S. (2015). Summary of the economic impacts of power quality on consumers. *2015 Australasia University Power Engineering Conference on Challenges of Future Grids, AUPEC 2015,* Wollongong, Australia (27–30 September 2015), 1–6. IEEE. https://ro.uow.edu.au/eispapers/4753/ (accessed 10 October 2022).

34 Rashed Mohassel, R., Fung, A., Mohammadi, F., and Raahemifar, K. (2014). A survey on advanced metering infrastructure. *International Journal of Electrical Power & Energy Systems* 63: 473–484.

35 IEEE Power and Energy Society (2019). *IEEE Recommended Practice for Power Quality Data Interchange Format (PQDIF)*. IEEE Std 1159.3-2019 (Revision of IEEE Std 1159.3-2003), 1–185. IEEE. https://doi.org/10.1109/IEEESTD.2019.8697192 (accessed 10 October 2022).

36 Sabin, D. and Dabbs, W.W. (2016). Paper 0258 on "Revision of IEEE STD 1159, 3 PQDIF." *23rd International Conference on Electricity Distribution,* Lyon (15–18 June 2015), 15–18. http://cired.net/publications/cired2015/papers/CIRED2015_0258_final.pdf (accessed 10 October 2022).

37 Saini, M.K. and Kapoor, R. (2012). Classification of power quality events – a review. *International Journal of Electrical Power & Energy Systems* 43 (1): 11–19. https://doi.org/10.1016/j.ijepes.2012.04.045.

38 Montoya, F.G., Manzano-Agugliaro, F., López, J.G., and Alguacil, P.S. (2012). Power quality research techniques: advantages and disadvantages. *DYNA* 79 (173, Part I): 66–74.

39 Kezunović, M., Rikalo, I., and Šobajić, D.J. (1995). High-speed fault detection and classification with neural nets. *Electric Power Systems Research* 34 (2): 109–116. https://doi.org/10.1016/0378-7796(95)00962-X.

40 Gu, I.Y.H. and Styvaktakis, E. (2003). Bridge the gap: signal processing for power quality applications. *Electric Power Systems Research* 66 (1): 83–96. https://doi.org/10.1016/S0378-7796(03)00074-9.

41 Gaouda, A.M., Salama, M.M.A., Sultan, M.R., and Chikhani, A.Y. (1999). Power quality detection and classification using wavelet-multiresolution signal decomposition. *IEEE Transactions on Power Delivery* 14 (4): 1469–1476. https://doi.org/10.1109/61.796242.

42 Santoso, S., Powers, E.J., Grady, W.M., and Hofmann, P. (1996). Power quality assessment via wavelet transform analysis. *IEEE Transactions on Power Delivery* 11 (2): 924–930. https://doi.org/10.1109/61.489353.

43 Xiangxun, C. (2002). Wavelet-based detection, localization, quantification and classification of short duration power quality disturbances. *Proceedings of IEEE Power Engineering Society Transmission and Distribution Conference* 2 (c): 931–936. https://doi.org/10.1109/pesw.2002.985142.

44 Ferreira, D.D., Marques, C.A.G., de Seixas, J.M. et al. (2011). Exploiting higher-order statistics information for power quality monitoring. *Power Quality*, vol. 2000. InTech https://doi.org/10.5772/14500.

45 Li, L. (2019). Analysis of uncertainty influence on the probabilistic evaluation of vortex-induced vibration response of a self-anchored suspension bridge. *KSCE Journal of Civil Engineering* 23 (11): 4790–4799. https://doi.org/10.1007/s12205-019-0208-x.

46 Troncossi, M. and Pesaresi, E. (2019). Analysis of synthesized non-Gaussian excitations for vibration-based fatigue life testing. *Journal of Physics Conference Series* 1264 (1): https://doi.org/10.1088/174 2-6596/1264/1/012039.

47 González-de-la-Rosa, J.J., Camarero, J.M., Bouaud S. et al. (2008). A higher-order statistics-based virtual instrument for termite activity targeting. *Fifth International Conference on Informatics in Control, Automation and Robotics – SPSMC*, 155–162. https://doi.org/ 10.5220/0001493701550162.

48 Bollen, M.H.J., Gu, I.Y.H., Santoso, S. et al. (2009). Bridging the gap between signal and power: assessing power system quality using signal processing techniques. *IEEE Signal Processing Magazine* 26 (4): 12–31. https://doi.org/10.1109/MSP.2009.932706.

49 Mendel, J.M. (1991). Tutorial on higher-order statistics (spectra) in signal processing and system theory: theoretical results and some applications. *Proceedings of the IEEE* 49 (3): 278–305. https://doi. org/10.1109/5.75086.

50 Ribeiro, M.V., Marques, C.A., Duque, C.A. et al. (2006). Power quality disturbances detection using HOS. *2006 IEEE Power Engineering Society, General Meeting, PES* 6p. https://doi.org/10.1109/pes.2006.1709131.

51 Ribeiro, M.V., Marques, C.A.G., Duque, C.A. et al. (2007). Detection of disturbances in voltage signals for power quality analysis using HOS. *EURASIP Journal of Advances in Signal Processing* 2007: 059786.

52 Ferreira, D.D., Cerqueira, A.S., Duque, C.A., and Ribeiro, M.V. (2009). HOS-based method for classification of power quality disturbances. *Electronics Letters* 45 (3): 183–185. https://doi.org/10.1049/el:20092969.

53 González-de-la-Rosa, J.J., Agüera-Pérez, A., Palomares-Salas, J.C., and Moreno-Muñoz, A. (2013). Higher-order statistics: discussion and interpretation. *Measurement Journal of International Measurement Confederation* 46 (8): 2816–2827. https://doi.org/10.1016/ j.measurement.2013.04.055.

54 Sierra-Fernández, J.M. (2017). Técnicas y procedimientos de medida basados en la Kurtosis Espectral. Una aplicación en el análisis de la calidad de la energía eléctrica. Tesis Doctoral. University of Cádiz.

55 González-de-La-Rosa, J.J., Moreno-Muñoz, A., and Puntonet, C.G. (2007). A practical approach to higher-order statistics. An application to electrical transients characterization. *2007 IEEE International Symposium on Intelligent Signal Processing, WISP,* Spain (3–5 October 2007), 224–229. IEEE.

56 Quirós-Olozábal, A., González-De-La-Rosa, J.J., Cifredo-Chacón, M.Á., and Sierra-Fernández, J.M. (2016). A novel FPGA-based system for real-time calculation of the spectral kurtosis: a prospective application to harmonic detection. *Measurement Journal of the International Measurement Confederation* 86: 101–113. https://doi.org/10.1016/j.measurement.2016.02.031.

57 Moreira, M.G., Ferreira, D.D., and Duque, C.A. (2016). Interharmonic detection and identification based on higher-order statistics. *Proceedings of the International Conference on Harmonic Quality Power, ICHQP.* 2016 (December): 679–684. https://doi.org/10.1109/ICHQP.2016.7783431.

58 Moreira, M.G., Ferreira, D.D., and Duque, C.A. (2014). Sub-harmonics detection and identification using higher order statistics. *Proceedings of the International Conference on Harmonic Quality Power, ICHQP,* Bucharest, Romania (25–28 May 2014), 283–287. IEEE Power and Energy Society.

59 González-de-la-Rosa, J.J., Sierra-Fernández, J.M., Agüera-Pérez, A. et al. (2013). An application of the spectral kurtosis to characterize power quality events. *International Journal of Electrical Power and Energy Systems* 49 (1): 386–398. https://doi.org/10.1016/j.ijepes.2013.02.002.

60 González-de-la-Rosa, J.J., Sierra-Fernández, J.M., Palomares-Salas, J.C. et al. (2015). An application of spectral kurtosis to separate hybrid power quality events. *Energies* 8 (9): 9777–9793. https://doi.org/10.3390/en8099777.

61 González-de-la-Rosa, J.J., Agüera-Pérez, A., Palomares-Salas, J.C. et al. (2012). A novel virtual instrument for power quality surveillance based in higher-order statistics and case-based reasoning. *Measurement* 45: 1824–1835. https://doi.org/10.1016/j.measurement.2012.03.036.

62 Sierra-Fernández, J.M., González-de-la-Rosa, J.J., Palomares-Salas, J.C. et al. (2012). HOS-based virtual instrument for power quality assessment. *IT Revolutions. Third International ICST Conference,* Cordoba, Spain. Part of the Lecture Notes of the Institute for Computer Sciences, Social Informatics and Telecommunications Engineering book series,Vol. 82, 1–9. https://doi.org/10.1007/978-3-642-32304-1_1.

63 González-de-la-Rosa, J.J. and Muñoz, A.M. (2008). Categorization of power quality transients using higher-order statistics and competitive layers-based neural networks. *CIMSA 2008 – IEEE Conference on Computer Intelligence Measurement System Application Procedures*, Istanbul, Turkey (14–16 July 2008), 83–86. IEEE Instrumentation and Measurement Society.

64 González-de-la-Rosa, J.J., Moreno-Muñoz, A., Palomares-Salas, J.C., and Agüera-Pérez, A. (2010). Automatic classification of power quality disturbances via higher-order cumulants and self-organizing networks. *IEEE International Symposium on Industrial Electronics*, Bari, Italy (4–7 July 2010), 1579–1584. IEEE.

65 Wang, J., Yen, G.G., and Polycarpou, M.M. (2012). *Lecture Advances in Neural Networks – ISNN Part II. The Recognition Study of Impulse and Oscillation Transient Based on Spectral Kurtosis and Neural Networks*, vol. 3535, 56–63. Berlin: Springer.

66 Li, W. and Liu, Z. (2015). An approach to classify transient disturbances with incomplete S-transform and Morlet wavelet spectral kurtosis. *2015 5th International Conference on Information Science Technology, ICIST*, Changsha, China (24–26 April 2015), Vol. 2015 (2), 249–253. IEEE. https://doi.org/10.1109/ICIST.2015.7288977 (accessed 10 October 2022).

67 Veena, V. and Kurian, A.A. (2014). Classification of power quality disturbances using time/frequency domain features. *2014 International Conference on Power Signals and Control Computing, EPSCICON* Thrissur, India (January 2014), 1–5. IEEE.

68 Lopez-Ramirez, M., Ledesma-Carrillo, L., Cabal-Yepez, E. et al. (2016). EMD-based feature extraction for power quality disturbance classification using moments. *Energies* 9 (7): 1–15. https://doi.org/10.3390/en9070565.

69 Florencias-Oliveros, O. (2020). Técnicas instrumentales para la monitorización de la calidad de la energía (instrumental techniques for power quality monitoring). Doctoral Thesis. University of Cádiz, https://www.educacion.gob.es/teseo/mostrarRef.do?ref=1843695 (accessed 06 February 2020).

70 Blanco, A.M., Gupta, M., Gil-de-Castro, A. et al. (2018). Impact of flat-top voltage waveform distortion on harmonic current emission and summation of electronic household appliances. *Renewable Energy and Power Quality Journal* 1 (April): 698–703.

71 Gil-de-Castro, A. (2012). Estudio y caracterización de la calidad de suministro eléctrico en los sistemas de alumbrado AUTOR. Tesis Doctoral. Universidad de Córdoba.

72 Kilter, J., Elphick, S., Meyer, J., and Milanović, J.V. (2014). Guidelines for power quality monitoring – results from CIGRE/CIRED JWG C4.112. *Proceedings from the International Conference on Harmonic Quality Power, ICHQP*, Bucharest, Romania (25–28 May 2014), 703–707. IEEE Power and Energy Society (PES).

73 Agüera-Pérez, A., Palomares-Salas, J.J., González-de-la-Rosa, J.J. et al. (2011). Characterization of electrical sags and swells using higher-order statistical estimators. *Measurement* 44: 1453–1460. https://doi.org/10.1016/j.measurement.2011.05.014.

74 Florencias-Oliveros, O., González-de-la-Rosa, J.J., Agüera-Pérez, A., and Palomares-Salas, J.C. (2018). Power quality event dynamics characterization via 2D trajectories using deviations of higher-order statistics. *Measurement Journal of the International Measurement Confederation* 125. https://doi.org/10.1016/j.measurement.2018.04.098.

75 Florencias-Oliveros, O., González-De-La-Rosa, J.J., Agüera-Pérez, A., and Palomares-Salas, J.C. (2019). Reliability monitoring based on higher-order statistics: a scalable proposal for the smart grid. *Energies* 12 (1). https://doi.org/10.3390/en12010055.

76 Bollen, M.H.J. and Hassan, F. (2011). Integration of distributed generation in the power system. *Integration of Distributed Generation in the Power System*. Chapters 1–8. IEEE Press Series on Power Engineering. Wiley.

77 Bollen, M.H.J. and Häger, M. (2005). Impact of increasing penetration of distributed generation on the number of voltage dips experienced by end-customers. *IEE Conference Publications* 4 (2005–11034): 255–259. https://doi.org/10.1049/cp:20051249.

78 Mohseni, M., Islam, S.M., and Masoum, M.A.S. (2011). Impacts of symmetrical and asymmetrical voltage sags on DFIG-based wind turbines considering phase-angle jump, voltage recovery, and sag parameters. *IEEE Transactions on Power Electronics* 26 (5): 1587–1598. https://doi.org/10.1109/TPEL.2010.2087771.

79 Agüera-Pérez, A., Palomares-Salas, J.C., González-de-la-Rosa, J.J., and Florencias-Oliveros, O. (2018). Weather forecasts for microgrid energy management: review, discussion and recommendations. *Applied Energy* 228, https://doi.org/10.1016/j.apenergy.2018.06.087.

80 International Electrotechnical Commission (2006). Electromagnetic compatibility (EMC) – Part 2–8: Environment – Voltage dips and short interruptions on public electric power supply systems with statistical measurement results.

81 Florencias-Oliveros, O., Aguera-Pérez, A., Sierra-Fernández, J.M. et al. (2018). Voltage supply frequency uncertainty influence on power quality index: a qualitative analysis of higher-order statistics 2D trajectories. *9th IEEE International Working Application Measurement Power Systems (AMPS) 2018 – Proceedings*, Bologna, Italy (September 26–28, 2018), 70–75. IEEE Instrumentation and Measurement Society (IMS).

82 IEEE Std 1588. (2019). *IEEE Standard for a Precision Clock Synchronization Protocol for Networked Measurement and Control Systems* (Revision of IEEE Std 1588-2008), 1–499. https://doi.org/10.1109/IEEESTD.2020.9120376.

83 IEEE Std 519. (2014). *IEEE Recommended Practice and Requirements for Harmonic Control in Electric Power Systems* (Revision of IEEE Std 519-1992), 1–29. https://doi.org/10.1109/IEEESTD.2014.6826459.

84 IEC 61000-4-7:2002/A1:2008. Part 4–7: Testing and measurement techniques – General guide on harmonics and interharmonics measurements and instrumentation, for power supply systems and equipment connected thereto.

85 González-de-la-Rosa, J.-J., Agüera-Pérez, A., Palomares-Salas, J.-C. et al. (2018). A dual monitoring technique for power quality transients based in the fourth-order spectrogram. *Energies* 11 (3): 503. https://doi.org/10.3390/en11030503.

86 González-de-la-Rosa, J.J., Agüera Pérez, A., Palomares Salas, J.C., and Sierra-Fernández, J.M. (2015). A novel measurement method for transient detection based in wavelets entropy and the spectral kurtosis: an application to vibrations and acoustic emission signals from termite activity. *Measurement* 68: 58–69. https://doi.org/10.1016/j.measurement.2015.02.044.

87 Sierra-Fernández, J.M., Rönnberg, S., González-de-la-Rosa, J.J. et al. (2019). Application of spectral kurtosis to characterize amplitude variability in power systems' harmonics. *Energies* 12 (1): 1–15. https://doi.org/10.3390/en12010194.

88 Ray, P.K., Foo Eddy, Y.S., Krishnan, A. et al. (2018). Wavelet transform-spectral kurtosis based hybrid technique for disturbance detection in a microgrid. *IEEE Power & Energy Society General Meeting* 2018. https://doi.org/10.1109/PESGM.2018.8586288.

89 Liu, Z., Zhang, Q., Han, Z., and Chen, G. (2014). A new classification method for transient power quality combining spectral kurtosis with neural network. *Neurocomputing* 125: 95–101. https://doi.org/10.1016/j.neucom.2012.09.037.

90 Eftekharnejad, B., Carrasco, M.R., Charnley, B., and Mba, D. (2011). The application of spectral kurtosis on acoustic emission and vibrations from a defective bearing. *Mechanical Systems and Signal Processing* 25 (1): 266–284. https://doi.org/10.1016/j.ymssp.2010.06.010.

91 Wang, Y. and Liang, M. (2011). An adaptive SK technique and its application for fault detection of rolling element bearings. *Mechanical Systems and Signal Processing* 25 (5): 1750–1764. https://doi.org/10.1016/j.ymssp.2010.12.008.

92 Guo, Y., Liu, T.W., Na, J., and Fung, R.F. (2012). Envelope order tracking for fault detection in rolling element bearings. *Journal of Sound and Vibrations* 331 (25): 5644–5654. https://doi.org/10.1016/j.jsv.2012.07.026.

93 Barszcz, T. and Randall, R.B. (2009). Application of spectral kurtosis for detection of a tooth crack in the planetary gear of a wind turbine. *Mechanical Systems and Signal Processing* 23 (4): 1352–1365. https://doi.org/10.1016/j.ymssp.2008.07.019.

94 Combet, F. and Gelman, L. (2009). Optimal filtering of gear signals for early damage detection based on the spectral kurtosis. *Mechanical Systems and Signal Processing* 23 (3): 652–668. https://doi.org/10.1016/j.ymssp.2008.08.002.

95 Liu, H., Huang, W., Wang, S., and Zhu, Z. (2014). Adaptive spectral kurtosis filtering based on Morlet wavelet and its application for signal transients detection. *Signal Processing* 96 (Part A): 118–124. https://doi.org/10.1016/j.sigpro.2013.05.013.

96 González-de-la-Rosa, J.J. and Moreno-Muñoz, A.M. (2008). Higher-order cumulants and spectral kurtosis for early detection of subterranean termites. *Mechanical Systems and Signal Processing* 22 (2): 279–294. https://doi.org/10.1016/j.ymssp.2007.08.009.

97 González-de-la-Rosa, J.J., Moreno-Muñoz, A., Gallego, A. et al. (2010). On-site non-destructive measurement of termite activity using the spectral kurtosis and the discrete wavelet transform. *Measurement* 43 (10): 1472–1488. https://doi.org/10.1016/j.measurement.2010.08.009.

98 González-de-la-Rosa, J.J., Agüera-Pérez, A., Palomares-Salas, J.C., and Sierra-Fernández, J.M. (2016). Wavelet filters and higher-order frequency analysis of acoustic emission signals from termite activity. *Measurement. Journal of the International Measurement Confederation* 93: 315–318. https://doi.org/10.1016/j.measurement.2016.07.037.

99 Søbjerg, S.S., Svoboda, J., Balling, J.E., and Skou, N. (2012). Detection of radio-frequency interference in microwave radiometers using spectral kurtosis. *IEEE International Geoscience and Remote Sensing Symposium* 1: 7141–7144. https://doi.org/10.1109/IGARSS.2012.6352016.

100 Awan, S.N., Krauss, A.R., and Herbst, C.T. (2015). An examination of the relationship between electroglottographic contact quotient, electroglottographic decontacting phase profile, and acoustical spectral moments. *Journal of Voice* 29 (5): 519–529. https://doi.org/10.1016/j.jvoice.2014.10.016.

101 Dion, J.L., Tawfiq, I., and Chevallier, G. (2012). Harmonic component detection: optimized spectral kurtosis for operational modal analysis. *Mechanical Systems and Signal Processing* 26 (1): 24–33. https://doi.org/10.1016/j.ymssp.2011.07.009.

102 Antoni, J. (2005). Blind separation of vibration components: principles and demonstrations. *Mechanical Systems and Signal Processing* 19 (6): 1166–1180. https://doi.org/10.1016/j.ymssp.2005.08.008.

103 Antoni, J. (2006). The spectral kurtosis: a useful tool for characterising non-stationary signals. *Mechanical Systems and Signal Processing* 20 (2): 282–307. https://doi.org/10.1016/j.ymssp.2004.09.001.

104 Collis, W.B., White, P.R., and Hammond, J.K. (1994). Higher order spectra and the trispectrum. *IEEE Digital Signal Processing Workshop*, Yosemite National Park, CA, (2–5 October 1994), 85–88. IEEE.

105 Collis, W.B., White, P.R., and Hammond, J.K. (1998). Higher-order spectra: the bispectrum and trispectrum. *Mechanical Systems and Signal Processing* 12 (3): 375–394. https://doi.org/10.1006/mssp.1997.0145.

106 Antoni, J. (2015). The spectral kurtosis of nonstationary signals: formalisation, some properties, and application. *European Signal Processing Conference* 06 (10 September): 1167–1170.

107 Dwyer, R.F. (2020). FRAM II Single Channel Ambient Noise Statistics: a Paper Presented at the 101st Meeting of the Acoustical Society of America, Ottawa, Canada (19 May 1981), 1–34. Naval Underwater Systems Center New London, CT New London Lab. https://apps.dtic.mil/sti/citations/ADA108755 (accessed 23 July 2022).

108 Braun, J., Gosbell, V., and Burnett, I. (2002). XML description schema for power quality data. *Proceedings of Australasia University Power Engineering Conference* (January 2003), 1–5. http://www.elec.uow.edu.au/apqrc/content/papers/AUPEC/AUPEC02_4.

109 Rönnberg, S.K. and Ackeby, S. (2015). Paper 0375, Very short variations in voltage (timescale less than 10 minutes) due to variations in wind and solar power. *23rd International Conference on Electricity Distribution*, Lyon (15–18 June), 15–18.

110 Domagk, M., Zyabkina, O., Meyer, J., and Schegner, P. (2015). Trend identification in power quality measurements. *2015 Australasia University Power Engineering Conference on Challenges in Future Grids*, AUPEC, Wollongong, Australia (27–30 September 2015). IEEE.

111 Domagk, M., Meyer, J., and Schegner, P. (2015). Seasonal variations in long-term measurements of power quality parameters. *2015 IEEE Eindhoven PowerTech Conference*, Eindhoven, Netherlands (June 29-July 02, 2015), 1–6.

112 Florencias-Oliveros, O., Agüera-Pérez, A., González-de-la-Rosa, J.J. et al. (2017). A novel instrument for power quality monitoring based in higher-order statistics: a dynamic triggering index for the smart grid.

Renewable Energy and Power Quality Journal 1 (15): 43–48. https://doi. org/10.24084/repqj15.212.

113 Florencias-Oliveros, O., González-De-La-Rosa, J.J., Sierra-Fernández, J.M. et al. (2022). Site characterization index for continuous power quality monitoring based on higher-order statistics. *Journal of Modern Power Systems and Clean Energy* 10 (1): 222–231. https://doi.org/10.35833/ MPCE.2020.000041.

114 Florencias-Oliveros, O., González-de-la-Rosa, J.J., Agüera-Pérez, A., and Palomares-Salas, J.C. (2018). Video-computational solutions for advanced metering infrastructure (AMI). Reliability monitoring based on higher-order statistics: a scalable proposal for the smart grid, ResearchGate. https://www. researchgate.net/publication/330933689_Video-Computational_ solutions_for_advanced_metering_infrastructure_AMI_Reliability_ Monitoring_Based_on_Higher-Order_Statistics_A_Scalable_Proposal_ for_the_Smart_Grid (accessed 10 October 2022).

115 Florencias-Oliveros, O., Espinosa-Gavira, M.J., González-dela-Rosa, J.J. et al. (2017). IEEE DataPort. Real-life power quality sags. *IEEE DataPort*. https://dx.doi.org/10.21227/H2K88D.

116 Florencias-Oliveros, O., Espinosa-Gavira, M.J., González-de-la-Rosa, J.J. et al. (2018). IEEE DataPort. Real-life power quality transients. *IEEE DataPort*. https://dx.doi.org/10.21227/H2Q30W.

Index

Power Quality Measurement and Analysis Using Higher-Order Statistics: Understanding HOS Contribution on the Smart(er) Grid, First Edition. Olivia Florencias-Oliveros, Juan-José González-de-la-Rosa, José-María Sierra-Fernández, Manuel-Jesús Espinosa-Gavira, Agustín Agüera-Pérez, and José-Carlos Palomares-Salas.
© 2023 John Wiley & Sons Ltd. Published 2023 by John Wiley & Sons Ltd.